Literature for Youth Series

Series Editor: Edward T. Sullivan

Discoveries and Inventions in Literature for Youth

A Guide and Resource Book

Joy L. Lowe
Kathryn I. Matthew

Literature for Youth, No. 3

The Scarecrow Press, Inc.
Lanham, Maryland, and Oxford
2004

SCARECROW PRESS, INC.

Published in the United States of America
by Scarecrow Press, Inc.
A wholly owned subsidary of
The Rowman & Littlefield Publishing Group, Inc.
4501 Forbes Boulevard, Suite 200, Lanham, Maryland 20706
www.scarecrowpress.com

PO Box 317
Oxford
OX2 9RU, UK

British Library Cataloguing in Publication Information Available

Library of Congress Cataloging-in-Publication Data

Lowe, Joy L.
  Discoveries and inventions in literature for youth : a guide and
resource book / Joy L. Lowe, Kathryn I. Matthew.
      p. cm. — (Literature for youth series ; no. 3)
  Includes bibliographical references and indexes.
  ISBN 0-8108-4915-1 (alk. paper)
  1. Children's literature—Bibliography. 2. Children--Books and
reading—United States—Bibliography. 3. Young adult
literature—Bibliography. 4. Teenagers—Books and reading—United
States—Bibliography. I. Matthew, Kathryn I. II. Title. III. Series.
Z1037 .L93 2004
011.62—dc22   2003022710

⊖™ The paper used in this publication meets the minimum requirements of
American National Standard for Information Sciences—Permanence of
Paper for Printed Library Materials, ANSI/NISO Z39.48-1992.
Manufactured in the United States of America.

For Chip and Lisa, Kira and Jim, and Michelle – J.L.L.

For Josh, Christie, Ben, and Erin – K.I.M.

Contents

Introduction

Inventions and discoveries include hidden dangers and face the scrutiny of skeptics who discount their value. Marie Curie's exposure to radium throughout her lifetime contributed to her death from leukemia, as the dangers of radiation were at that time unknown. The Wright brothers faced skeptics who were certain that objects heavier than air could not fly. Dyson (2001) contends that knowing when to ignore the experts is important. Intrepid inventors and discoverers persist despite the dangers and the skeptics.

Many inventors start by seeking a solution to a personal problem. For example, Madam C. J. Walker's hair loss led to her experiments to find a homemade remedy. Her successful results formed the basis of her hair care product empire. With persistence, dedication, and hard work she began to make and market hair care products, eventually becoming the first female African American self-made millionaire. At a time when few women and few African Americans were successfully inventing and marketing their inventions, Madam C. J. Walker did.

Ben Franklin communicated with other scientists who were experimenting with electricity. They shared ideas, successes, and challenges. Franklin was dismayed to discover that some of his experiments were not taken seriously and ended up as parlor games. While he recognized the power and potential of electricity he could not have foreseen how dependent modern society would become on electricity.

Learning about discoveries and inventions satisfies students' curiosity about the origins of products they use every day. More importantly, they learn the value of hard work, persistence, and working with others to achieve goals. They learn to look for solutions to problems as they develop their creative and critical thinking skills. Students develop wariness about the potential advantages and disadvantages of discoveries and inventions. They learn to look at new things with an open mind, to be aware of their possibilities, to not dismiss them outright, but to proceed with caution as they accept new discoveries and inventions.

The purpose of this book is to provide librarians and teachers with a rich, varied collection of books and other resources on discoveries and inventions. Chapters in the book contain annotations of selected books and other resources appropriate for kindergarten through twelfth grade as they learn about discoveries and inventions. The annotations in each chapter are listed first by grade levels and within grade levels; they are arranged in alphabetical order by the author's last name.

Chapter 1 includes biographies and autobiographies of individual discoverers and inventors whose hard work, research, observations, and perseverance have led to significant achievements.

Chapter 2 contains collective biographies that profile a group of discoverers or inventors such as women in science, African American inventors, Hispanic scientists, and Asian American scientists.

Chapter 3 has annotations for picture books from different genres. These are books about transportation, inventions, discoveries, inventors, and time travel.

Chapter 4 includes fiction titles for reading for pleasure. Included in this chapter are selected books of poetry, fantasy, science fiction, and realistic fiction.

Chapter 5 has nonfiction books about inventions that are part of our everyday lives, such as the clock, the sewing machine, and the camera.

Chapter 6 focuses on nonfiction books about industry, energy, and the environment. Included in this chapter are books on the Industrial Revolution, construction, different forms of energy, and efforts to save the environment.

Chapter 7 includes nonfiction books on information technology, which encompasses computers, the Internet, calculating machines, and virtual reality.

Chapter 8 contains nonfiction books about media and communications including writing, the telegraph, the radio, the telephone, and the television.

Chapter 9 has annotations for nonfiction books on medicine and health including books on vaccines, anesthesia, biotechnology, and infectious diseases.

Chapter 10 contains annotations for nonfiction science books on subjects such as simple machines, physics, the scientific revolution, microscopes, and other scientific discoveries and inventions.

Chapter 11 includes nonfiction books on space, which focuses on, among other topics, discoveries in space, inventions that make space travel possible, and satellites.

Chapter 12 contains annotations for nonfiction books about different forms of transportation including airplanes, trains, and automobiles.

Chapter 13 looks at nonfiction books on weapons and warfare such as remote-controlled weapons, radar, and other tools of war.

Chapter 14 has annotations of nonfiction books that contain information on collections of inventions such as great inventions, inventions that changed the world, and inventions that happened by accident.

Chapter 15 looks at nonfiction books that describe how things work. Books in this collection take readers inside inventions such as dolls that cry tears, cameras, and rockets.

Chapter 16 looks to the future with nonfiction books on robots, bionics, time machines, and artificial intelligence.

Chapter 17 contains annotations on reference books including dictionaries, encyclopedias, and timelines.

Chapter 18 provides annotations for books to help would-be inventors design, produce, patent, and market their inventions.

Appendix A is a collection of professional resources for teachers and librarians. Within this collection are book annotations, periodicals, articles, and information on invention contests for students to enter.

Appendix B contains information on electronic resources, which includes software, videos, DVDs, and Internet websites for gathering additional information.

Appendix C is a directory of museums, organizations, and other institutions. Contact information is provided as well as websites.

Appendix D is a collection of booktalks, classroom activities, and invention contests.

Appendix E contains a suggested list of books for a core library curriculum organized by grade levels.

References

Dyson, James. *A History of Great Inventions*. New York: Carroll and Graf, 2001.

Chapter 1

Biographies and Autobiographies

These are the biographies and autobiographies of discoverers, inventors, scientists, and thinkers who, with hard work, perseverance, courage, and in some cases serendipity, have made a profound impact on our world from mountains to microbes and from space to the deep blue sea. Reading biographies and autobiographies helps children realize that they too can contribute to making the world a better place.

Fritz, Jean. *What's the Big Idea, Ben Franklin?* Illus. by Margot Tomes. New York: Putnam, 1976. 48 pp. (hc. 0-399-23487-X) $16.99; (pbk. 0-698-11372-1) $7.95.
Gr. 1–3. Benjamin Franklin's curiosity and genius led him to explore and discover his world. He questioned things around him and worked to find answers and solve problems. One of his discoveries was that electricity and lightning are the same.

McLoone, Margo. *George Washington Carver*. Mankato, Minn.: Capstone Press, 1997. 24 pp. (hc. 1-56065-516-X) $18.60.
Gr. 1–3. From the Read and Discover Photo-Illustrated Biographies series. This brief biography introduces young readers to a distinguished scientist, who inspired others with his love of nature, helped farmers raise better crops, and introduced the world to a variety of products he created from peanuts and sweet potatoes. Photographs, quotes from George Washington Carver, a chronology, a glossary, resources for learning more, and an index are included in the book.

Sherrow, Victoria. *Alexander Graham Bell.* Illus. by Elaine Verstraete.
    Minneapolis: Carolrhoda Books, 2001. 48 pp. (hc. 1-57505-460-4)
    $22.60; (pbk. 1-57505-533-3) $5.95.
Gr. 1–3. From the On My Own Biography series. Readers learn about
Bell's early years working with his father on Visible Speech, his work
with the deaf, and his curious inventive mind. Sherrow helps readers
understand how these things influenced his life and his work. Colorful
illustrations, large print, and easy to read text make this biography accessible to young readers. The book concludes with an afterword and a
timeline.

Ford, Carin T. *Alexander Graham Bell: Inventor of the Telephone.*
    Berkeley Heights, N.J.: Enslow, 2002. 32 pp. (hc. 0-7660-1858-X)
    $17.95.
Gr. 2–4. From the Famous Inventors series. Born in Scotland, Bell was
interested in discovery as a small child. His first invention was a machine to separate wheat from its husks, made for the father of a friend.
As he grew older, he became interested in speech and hearing. Although he will always be known for his invention of the telephone, he
was proudest of his work with the deaf. This book concludes with a
timeline, words to know, books from which to learn more, Internet addresses, and an index.

——. *Thomas Edison: Inventor.* Berkeley Heights, N.J.: Enslow, 2002.
    32 pp. (hc. 0-7660-1860-1) $17.95.
Gr. 2–4. From the Famous Inventors series. Thomas Edison was an
extremely curious child. He asked questions of everyone all day long.
He loved to mix chemicals together just to see what would happen,
frequently with disastrous results! As a man, he created more than a
thousand inventions, such as the phonograph, motion pictures, and a
storage battery. He is best known for the discovery of and use of electricity. At the end of this book are a timeline, words to know, books
from which to learn more, Internet addresses, and an index.

——. *Benjamin Franklin: Inventor and Patriot.* Berkeley Heights, N.J.:
    Enslow, 2003. 32 pp. (hc. 0-7660-1859-8) $17.95.
Gr. 2–4. From the Famous Inventors series. Probably one of the most
inventive and accomplished statesmen of all time was Benjamin Franklin. Born and reared in Boston, Massachusetts, he apprenticed to his
older brother in his newspaper enterprise for a time. He was never
content to do only one thing and began a number of practical endeavors

including inventing. He finally attached himself to the political world and was powerful during the American Revolution. Included in this book are a timeline, a glossary, book titles, Internet addresses, and an index.

Hinman, Bonnie. *Benjamin Banneker: American Mathematician and Astronomer*. Philadelphia: Chelsea House Publishers, 2000. 79 pp. (pbk. 0-7910-5691-0) $8.95.
Gr. 2–4. From the Colonial Leaders series. Benjamin Banneker was a free African American whose curiosity, eagerness to learn, and willingness to work hard enabled him to write an almanac, build a clock, and survey the land that would become America's capital. Captioned illustrations, photographs, and reproductions enhance the text. The book includes a glossary, a chronology, a timeline of Colonial America, books for further reading, and an index.

Fisher, Leonard Everett. *Marie Curie*. New York: Macmillan Publishing, 1994. 27 pp. (hc. 0-02-735375-3) $14.95.
Gr. 2–5. Dark, foreboding black-and-white illustrations chronicle the life of this brilliant woman scientist who dedicated her life to her research. She fought against sexism and prejudice throughout her life and was awarded Nobel Prizes in both physics and chemistry. The compelling text begins with a chronology of her life and ends with additional information about her life.

Old, Wendie. *To Fly: The Story of the Wright Brothers*. Illus. by Robert Andrew Parker. Boston: Houghton Mifflin, 2002. 48 pp. (hc. 0-618-13347-X) $16.00.
Gr. 2–5. Old shows readers how the two brothers' personalities complemented each other and how they worked together to invent the airplane. The watercolor illustrations capture the tone of the text and the anticipation that something wonderful is going to happen. The book concludes with an epilogue, a flight timeline, books for further reading, notes, and an index.

Cefrey, Holly. *The Inventions of Alexander Graham Bell: The Telephone*. New York: PowerKids Press, 2003. 24 pp. (hc. 0-8239-6441-8) $17.25.
Gr. 2–6. From the Nineteenth Century Inventors series. While remembered for his invention of the telephone, Alexander Graham Bell, also

spent his life working with the hearing impaired. Books in this series contain high-interest facts about the inventors, illustrations, and large print making the material accessible to challenged readers. A glossary, resources for learning more, and an index are included.

——. *The Inventions of Martha Coston: Signal Flares That Save Sailors' Lives*. New York: PowerKids Press, 2003. 24 pp. (hc. 0-8239-6442-6) $17.25.
Gr. 2–6. From the Nineteenth Century Inventors series. Here is the story of how Coston perfected and marketed her signal flares. Books in this series contain high-interest facts about the inventors, illustrations, and large print making the material accessible to challenged readers. A glossary, resources for learning more, and an index are included.

——. *The Inventions of Thomas Alva Edison: Father of the Light Bulb and the Motion Picture Camera*. New York: PowerKids Press, 2003. 24 pp. (hc. 0-8239-6440-X) $17.25.
Gr. 2–6. From the Nineteenth Century Inventors series. This is a brief biography of Edison and it includes information on some of his more famous inventions. Books in this series contain high-interest facts about the inventors, illustrations, and large print making the material accessible to challenged readers. A glossary, resources for learning more, and an index are included.

——. *The Inventions of Amanda Jones: The Vacuum Method of Canning and Food Preservation*. New York: PowerKids Press, 2003. 24 pp. (hc. 0-8239-6445-0) $17.25.
Gr. 2–6. From the Nineteenth Century Inventors series. This is the story of Amanda Jones who devised better ways to can fruits and vegetables. Books in this series contain high-interest facts about the inventors, illustrations, and large print making the material accessible to challenged readers. A glossary, resources for learning more, and an index are included.

——. *The Inventions of Eli Whitney: The Cotton Gin*. New York: PowerKids Press, 2003. 24 pp. (hc. 0-8239-6443-4) $17.25.
Gr. 2–6. From the Nineteenth Century Inventors series. Readers learn about Whitney, his cotton gin, and its impact on the economy. Books in this series contain high-interest facts about the inventors, illustrations, and large print making the material accessible to challenged readers. A glossary, resources for learning more, and an index are included.

———. *The Inventions of Granville Woods: The Railroad Telegraph System and the "Third Rail."* New York: PowerKids Press, 2003. 24 pp. (hc. 0-8239-6442-6) $17.25.
Gr. 2–6. From the Nineteenth Century Inventors series. At the time of his death Woods held over sixty patents, many of which made railroad travel safer. His "third rail" provided more electricity to trains and is still in use today. Books in this series contain high-interest facts about the inventors, illustrations, and large print making the material accessible to challenged readers. A glossary, resources for learning more, and an index are included.

McKissack, Patricia, and Fredrick McKissack. *Madam C. J. Walker: Self-Made Millionaire.* Berkeley Heights, N.J.: Enslow, 1994. 128 pp. (hc. 0-76601-628-X) $14.95.
Gr. 3–4. From the Great African Americans series. After the death of her husband, Sarah Breedlove supported herself and her daughter by taking in laundry. However, Sarah wanted to better herself. She believed that African American women would feel better about themselves if they looked better, so she began a line of beauty products, which she sold door to door. She became the first female self-made millionaire in America. Her second husband was Charles Joseph Walker, and Sarah became known as Madam C. J. Walker. A glossary with brief, clear definitions and a short index are included.

Alphin, Elaine Marie. *Germ Hunter: A Story about Louis Pasteur.* Illus. by Elaine Verstraete. Minneapolis: Carolrhoda Books, 2003. 64 pp. (hc. 1-57505-179-6) $21.27.
Gr. 3–5. From the Creative Minds Biography series. This bibliography reads like a novel and is likely to appeal to reluctant readers. The focus of the biography is on Pasteur's life rather than his work. A glossary, a bibliography, and an index are included.

Bowen, Andy Russell. *A Head Full of Notions: A Story about Robert Fulton.* Minneapolis: Carolrhoda Books, 1997. 64 pp. (hc. 0-87614-876-3) $21.27; (pbk. 1-57505-026-9) $5.95.
Gr. 3–5. From a Carolrhoda Creative Minds Book series. Robert Fulton had a knack for looking at things and inventing ways to improve them. One example of this was his improvements to the steamboat that made it commercially viable. Illustrations, a bibliography, and an index are included.

Brackett, Virginia. *Steve Jobs: Computer Genius of Apple.* Berkeley
    Heights, N.J.: Enslow, 2003. 48 pp. (hc. 0-7660-1970-5) $18.95.
Gr. 3–5. From the Internet Biographies series. Steve Jobs was an unusual young man from an early age. He had an insatiable curiosity but
was very uncomfortable in school. So smart that his teachers suggested
that he skip a grade, he was pretty much a "loner." He always seemed
to "think outside the box" and connected with a Hewlett-Packard em-
ployee to form a company to design the world's first "personal computer," which they called the Apple. This book includes a chronology,
chapter notes, a glossary, books for further reading, Internet addresses,
and an index.

Brallier, Jess. *Who Was Albert Einstein?* Illus. by Robert Andrew
    Parker. New York: Grosset and Dunlap, 2002. 105 pp. (hc. 0-448-
    42659-5) $13.89; (pbk. 0-448-42496-7) $4.99.
Gr. 3–5. From the Who Was? series. Unruly hair, rumpled clothes, and
a penchant for thinking are one way to describe this genius. The book
combines an enjoyable text with black and white line drawings and just
the right tone to get young readers interested in reading to learn about
Einstein's impact on the world. The author brings up the question of the
contributions of Einstein's first wife, Mileva, to his theories. Timelines
of his life and the world during his lifetime conclude the book.

Burch, Joann Johansen. *Fine Print: A Story about Johann Gutenberg.*
    Illus. by Kent Alan Aldrich. Minneapolis: Carolrhoda Books,
    1991. 64 pp. (hc. 0-87614-682-5) $21.27.
Gr. 3–5. From the Creative Minds Biography series. Black and white
woodcuts illustrate this brief biography of the man who invented the
printing press. The description of Gutenberg's struggles to make a
working printing press is particularly engrossing. An author's note and
a glossary begin the book and a bibliography concludes the book.

Fisher, Leonard Everett. *Alexander Graham Bell.* New York:
    Atheneum, 1999. 32 pp. (hc. 0-689-81607-3) $16.00.
Gr. 3–5. Not only was Bell a prolific inventor he was also a humanitarian and an educator, as evidenced in part by his work with the deaf.
Readers learn about several of Bell's inventions and the impact of the
inventions on daily life. Black and white acrylic illustrations are included as is a chronology of Bell's life; however, the book lacks a bibliography.

Garty, Judy. *Jeff Bezos: Business Genius of Amazon.com.* Berkeley Heights, N.J.: Enslow, 2003. 48 pp. (hc. 0-7660-1972-1) $18.95.
Gr. 3–5. From the Internet Biographies series. In 1995, Jeff Bezos opened what he called "the biggest bookstore in the world." More than a "place," this bookstore was an "idea." Bezos's store was on the Internet at Amazon.com. He has become an awarded businessperson and a multi-billionaire. He has a motto: "work hard, have fun, make history." Ending the book are a chronology, chapter notes, a glossary, books for further reading, Internet addresses, and an index.

Giblin, James Cross. *The Amazing Life of Benjamin Franklin.* Illus. by Michael Dooling. 48 pp. New York: Scholastic, 2000. (hc. 0-590-48534-2) $17.95.
Gr. 3–5. This noted statesman was committed to helping the colonies become free from English rule and to assuring that the new nation succeeded. Some of the things he is remembered for include establishing libraries, writing *Poor Richard's Almanack*, and experimenting with electricity. Included in the book is a chronology of his life, a list of his inventions, a list of historic sites, excerpts from his *Almanack*, a bibliography, an artist's note, and an index.

Kramer, Barbara. *George Washington Carver: Scientist and Inventor.* Berkeley Heights, N.J.: Enslow, 2002. 128 pp. (hc. 0-7660-1770-2) $20.95.
Gr. 3–5. From the African American Biographies series. Born in Missouri in the 1860s, George Carver decided later that he wanted a middle name and chose "Washington." He was always fascinated by science and eventually became a teacher at the famed Tuskegee Institute. He became the farmer's best friend and developed the idea of crop rotation to replenish the soil. He also developed more than three hundred peanut-based products. This book includes a chronology, selected recipes by Dr. Carver, chapter notes, further reading and Internet addresses, and an index.

Lantier, Patricia. *Louis Braille.* Adapted by Beverly Birch. Milwaukee: Gareth Stevens Children's Books, 1991. 68 pp. (hc. 0-8368-0454-6) $21.26.
Gr. 3–5. From the People Who Made a Difference series. Blinded by an accident in his father's workshop at the age of three, Louis Braille is remembered for the development of the Braille alphabet that enables

the blind to read with their fingertips. Resources for learning more, a glossary, a timeline, a map, and an index conclude the book.

Morales, Leslie. *Esther Dyson: Internet Visionary*. Berkeley Heights, N.J.: Enslow, 2003. 48 pp. (hc. 0-7660-1973-X) $18.95.
Gr. 3–5. From the Internet Biographies series. Esther Dyson was born in Switzerland to a British science writer-astrophysicist and a mathematician. Her thinking led her to ideas instead of products and her company holds conferences and seminars for entrepreneurs, technology company executives, and investors to meet and exchange ideas. This book includes a chronology, chapter notes, a glossary, books for further reading, Internet addresses, and an index.

Peters, Craig. *Steve Case: Internet Genius of America Online*. Berkeley Heights, N.J.: Enslow, 2003. 48 pp. (hc. 0-7660-1971-3) $18.95.
Gr. 3–5. From the Internet Biographies series. As a youngster, Steve Case was an entrepreneur. He and his brother sold limeade with the limes they grew in their backyard in Hawaii, delivered newspapers, and sold seeds and greeting cards door-to-door. Steve always looked for a challenge and after trying jobs at Proctor and Gamble and Pizza Hut he became enamored with the computer business. He established America Online and later bought Time Warner, the largest entertainment conglomeration in the world. The book concludes with a chronology, chapter notes, a glossary, books for further reading, Internet addresses, and an index.

——. *Larry Ellison: Database Genius of Oracle*. Berkeley Heights, N.J.: Enslow, 2003. 48 pp. (hc. 0-7660-1974-8) $18.95.
Gr. 3–5. From the Internet Biographies series. Unlike many of his contemporaries in the computer field, Larry Ellison never took a computer class in his life, nor did he ever graduate from college. He grew up in a Chicago ghetto and in college he picked up a book and taught himself computer programming. In 1977, he co-founded Oracle, a software company that was second only to Microsoft and he became one of the richest men in the world. This book includes a chronology, chapter notes, a glossary, books for further reading, Internet addresses, and an index.

——. *Bill Gates: Software Genius of Microsoft*. Berkeley Heights, N.J.: Enslow, 2003. 48 pp. (hc. 0-7660-1969-1) $18.95.

Gr. 3–5. From the Internet Biographies series. At the age of thirteen, Bill Gates wrote his first computer program. He grew up to transform American business and to become the youngest billionaire the world has ever known. Gates is known not only for his acumen but also for his generosity. He and his wife formed a foundation that donates millions of dollars around the globe. His life story is as fascinating as it is inspiring. The book concludes with a chronology, chapter notes, a glossary, books for further reading, Internet addresses, and an index.

Gherman, Beverly. *The Mysterious Rays of Dr. Roentgen.* Illus. by Stephen Marchesi. New York: Atheneum, 1994. 24 pp. (hc. 0-689-31839-1) $14.95.
Gr. 3–6. This brief chapter book with colorful illustrations introduces readers to Roentgen's work with x-rays. Readers learn how early users of x-rays discovered the dangers of the mysterious rays along with the benefits. The book concludes with an author's note, a timeline, and an index.

Gourley, Catherine. *Wheels of Time: A Biography of Henry Ford.* Brookfield, Conn.: The Millbrook Press, 1997. 48 pp. (hc. 0-7613-0214-X) $21.90.
Gr. 3–6. Quotes from Henry Ford, photographs, reproductions, and etchings add interest to this biography. Working in conjunction with the Henry Ford Museum, the author sketches an inspiring tale of the man responsible for producing reliable, affordable automobiles. A bibliography concludes the book.

Matthews, Tom L. *Always Inventing: A Photobiography of Alexander Graham Bell.* Washington, D.C.: National Geographic Society, 1999. 64 pp. (hc. 0-7922-7391-5) $16.95.
Gr. 3–6. From the National Geographic Photobiography series. It is fitting that the National Geographic Society published this book, as Bell was one of the original members of the Society. Full-page photographs and illustrations depict his life and inventions. Best remembered for his invention of the telephone, Bell spent his life inventing. The book includes a chronology, a bibliography, some websites, and an index.

Maupin, Melissa. *Benjamin Banneker.* Chanhassen, Minn.: The Child's World, 2000. 40 pp. (hc. 1-56766-618-3) $25.64.

Gr. 3–6. This biography begins with a brief introduction to Banneker's grandparents and parents. He is considered to be the first African American scientist. His supportive, loving family encouraged the bright, capable, curious boy to attend school and excel. Captioned illustrations, a timeline, a glossary, an index, and resources for further research are included.

Sullivan, George. *The Wright Brothers*. New York: Scholastic, 2003. 128 pp. (pbk. 0-439-26320-4) $4.50.
Gr. 3–6. From the In Their Own Words series. The Wright brothers' letters, journal entries, notes, interviews, and diagrams help to tell the story of these aviation pioneers' adventures. The book concludes with a chronology, a bibliography, books for further reading, and an index.

Whitehurst, Susan. *Dr. Charles Drew, Medical Pioneer*. Chanhassen, Minn.: The Child's World, 2002. 40 pp. (hc. 1-56766-916-6) $28.50.
Gr. 3–6. Known as "the father of the blood bank," Drew is also remembered for inspiring other African Americans to overcome barriers, face challenges, and become doctors. Photographs, a timeline, a glossary, an index, and resources for further information are included.

Naden, Corinne J., and Rose Blue. *Jonas Salk: Polio Pioneer*. Brookfield, Conn.: The Millbrook Press, 2001. 48 pp. (hc. 0-7613-1804-6) $23.90.
Gr. 3–8. From A Gateway Biography series. The focus of this biography is on Salk's work to create a polio vaccine. The book contains background information to help readers understand the importance of the polio vaccine and of the controversy surrounding his killed virus vaccine and Sabin's live virus vaccine. When he died, Salk was working on an experimental AIDS vaccine. Photographs, a chronology, a listing of books and websites for learning more, and an index are included.

Parker, Steve. *Aristotle and Scientific Thought*. Broomall, Pa.: Chelsea House, 1995. 32 pp. (hc. 0-7910-3005-9) $18.55.
Gr. 3–8. From the Science Discoveries series. Young readers are surprised to discover that this ancient Greek philosopher's influence is evident today in the scientific world. Color illustrations, a timeline, a glossary, and an index are included.

——. *Alexander Graham Bell and the Telephone*. Broomall, Pa.: Chelsea House, 1995. 32 pp. (hc. 0-7910-3004-0) $18.55.
Gr. 3–8. From the Science Discoveries series. The telephone is an everyday object that student's value and in this biography, they learn a great deal about its inventor. Color illustrations, a timeline of the world during Bell's life, a glossary, and an index are included.

——. *Marie Curie and Radium*. Broomall, Pa.: Chelsea House, 1995. 32 pp. (hc. 0-7910-3011-3) $18.55.
Gr. 3–8. From the Science Discoveries series. Curie's discovery of radium and her experimentation with radioactive elements provided the information used to move the world into the Atomic Age. Photographs, reproductions, and sidebars enhance this informative text. The book includes a timeline of the world during Curie's time, a glossary, and an index.

——. *Charles Darwin and Evolution*. Broomall, Pa.: Chelsea House, 1995. 32 pp. (hc. 0-7910-3007-5) $18.55.
Gr. 3–8. From the Science Discoveries series. Readers discover the revolutionary work of Darwin and his theory of evolution in this engrossing biography. At the end of the book, his work is connected to the work of Gregor Mendel to help students understand the theory of evolution. Color illustrations, a timeline of the world during Darwin's life, a glossary, and an index are included.

——. *Thomas Edison and Electricity*. Broomall, Pa.: Chelsea House, 1995. 32 pp. (hc. 0-7910-3012-1) $18.55.
Gr. 3–8. From the Science Discoveries series. When Edison had problems at school, his mother decided to educate him at home. He spent his time reading scientific books and conducting experiments. His scientific experiments and his entrepreneurial ventures provided the foundation he needed to become an inventor and a businessman. Photographs, diagrams, reproductions, and sidebars complement the text and draw readers into the book. It concludes with a timeline of the world during Edison's life, a glossary, and an index.

——. *Albert Einstein and Relativity*. Broomall, Pa.: Chelsea House, 1995. 32 pp. (hc. 0-7910-3003-2) $18.55.
Gr. 3–8. From the Science Discoveries series. The genius of Einstein and his revolutionary work in physics are explored in this concise biog-

raphy. Color illustrations, a timeline of the world during Einstein's life, a glossary, and an index are included.

——. *Benjamin Franklin and Electricity.* Broomall, Pa.: Chelsea House, 1995. 32 pp. (hc. 0-7910-3006-7) $18.55.
Gr. 3–8. From the Science Discoveries series. Students familiar with Benjamin Franklin's contributions to Colonial America may not realize that he is also remembered for his experiments with electricity and numerous inventions. Color illustrations, a timeline of the world during Franklin's life, a glossary, and an index are included.

——. *Galileo and the Universe.* Broomall, Pa.: Chelsea House, 1995. 32 pp. (hc. 0-7910-3008-3) $18.55.
Gr. 3–8. From the Science Discoveries series. Galileo's inventions and discoveries were not accepted during his lifetime and caused him to be imprisoned as a heretic. Illustrations, photographs, diagrams, and sidebars provide intriguing details about the life and work of this mathematician, physicist, and astronomer. At the end of the book readers learn how Isaac Newton built on Galileo's work. The book concludes with a timeline of the world during Galileo's life, a glossary, and an index.

——. *Marconi and Radio.* Broomall, Pa.: Chelsea House, 1995. 32 pp. (hc. 0-7910-3009-1) $18.55.
Gr. 3–8. From the Science Discoveries series. Students familiar with wireless telephones and wireless computer networks enjoy reading about the inventor of the radio, our first means of wireless communication. Color illustrations, timeline of the world during Marconi's life, a glossary, and an index are included.

——. *Isaac Newton and Gravity.* Broomall, Pa.: Chelsea House, 1995. 32 pp. (hc. 0-7910-3010-5) $18.55.
Gr. 3–8. From the Science Discoveries series. In this biography readers learn that Newton, while noted for his study of gravity, also studied light, color, force, and motion. Color illustrations, a timeline of the world during Newton's life, a glossary, and an index are included.

——. *Louis Pasteur and Germs.* Broomall, Pa.: Chelsea House, 1995. 32 pp. (hc. 0-7910-3002-4) $18.55.
Gr. 3–8. From the Science Discoveries series. Louis Pasteur's contributions include founding the science of microbiology and developing the process of pasteurization. The book concludes with information on

terest was in inventing. His invention of the artificial heart was spurred
on by his desire to assist those suffering from heart disease, which
killed his father. The book concludes with a chronology of Jarvik's life,
an artificial heart timeline, resources for further reading, a glossary, and
an index.

——. *Jonas Salk and the Polio Vaccine*. Bear, Del.: Mitchell Lane,
      2002. 48 pp. (hc. 1-58415-093-9) $17.95.
Gr. 4–8. From the Unlocking the Secrets of Science series. This biography provides a brief, informative examination of Salk's work to develop the polio vaccine. Readers learn about the history of polio, Albert
Sabin's polio vaccine, and the other researchers whose efforts contributed to the creation of the vaccine. A chronology, a timeline, resources
for further reading, a glossary, and an index are included.

——. *Edward Teller and the Development of the Hydrogen Bomb*. Bear,
      Del.: Mitchell Lane, 2003. 48 pp. (hc. 1-58415-108-0) $17.95.
Gr. 4–8. From the Unlocking the Secrets of Science series. Teller is
known as "the father of the hydrogen bomb." He was hailed both as a
hero and as a demon for developing the hydrogen bomb. This biography includes information on the work of scientists whose discoveries
made the hydrogen bomb possible. A chronology, a timeline, resources
for further reading, a glossary, and an index conclude the book.

Blashfield, Jean F. *Carl Sagan: Astronomer*. Chicago: Ferguson Pub-
      lishing, 2001. 127 pp. (hc. 0-89434-374-2) $16.95.
Gr. 4–8. From the Ferguson Career Biographies series. Sagan has often
been denied recognition for his work as an astronomer because some do
not consider him a "real" scientist. He spent a great deal of his time
explaining the universe to others by writing and appearing on television. Black and white photographs, a timeline, information about becoming an astronomer, resources for learning more about astronomers
and Carl Sagan, and an index are contained in the book.

Fox, Mary Virginia. *Edwin Hubble: American Astronomer*. New York:
      Franklin Watts, 1997. 112 pp. (hc. 0-531-11371-X) $20.00.
Gr. 4–8. From A Book Report Biography series. On his eighth birthday, Hubble spent the night looking at the stars through his grandfather's telescope. This was just the first of many nights he would spend
searching the night sky for answers to questions about the universe.

Black and white photographs coupled with intriguing details tell the story of Hubble's life work. A chronology, notes on sources, resources for learning more, and an index conclude the book.

Freedman, Russell. *The Wright Brothers: How They Invented the Airplane*. New York: Holiday House, 1991. 129 pp. (hc. 0-8234-0875-2) $19.95; (pbk. 0-8234-1082-X) $12.95.
Gr. 4–8. Lavishly illustrated with photographs, some taken by Wilbur and Orville, this book chronicles their years of hard work and collaboration. Excerpts from primary sources such as journals and letters enhance the carefully crafted text. The book concludes with places to visit, books for further reading, and an index. This is a Newbery Honor book.

Gaines, Ann. *Tim Berners-Lee and the Development of the World Wide Web*. Bear, Del.: Mitchell Lane, 2002. 48 pp. (hc. 1-58415-095-5) $17.95.
Gr. 4–8. From the Unlocking the Secrets of Science series. The author describes why and how the Internet was developed. Then readers learn how Berners-Lee developed the World Wide Web and why he gave it to the public rather than trying to make money from the Web. A Tim Berners-Lee chronology, a World Wide Web timeline, resources for further reading, a glossary, and an index are included.

——. *Wallace Carothers and the Story of Dupont Nylon*. Bear, Del.: Mitchell Lane, 2002. 48 pp. (hc. 1-58415-097-1) $17.95.
Gr. 4–8. From the Unlocking the Secrets of Science series. A shy, quiet man, Carothers enjoyed conducting pure research in his chemistry laboratory. In 1827, he was hired by the DuPont Company to do just that without pressure to produce a viable product. His research and the research of his associates eventually led to the development of commercially viable substances including neoprene and nylon. Throughout his life he suffered bouts of depression and at the age of forty-one he committed suicide. A Wallace Carothers chronology, a nylon timeline, resources for further reading, a glossary, and an index are included.

Gaines, Ann, and Jim Whiting. *Robert A. Weinberg and the Search for the Cause of Cancer*. Bear, Del.: Mitchell Lane, 2002. 48 pp. (hc. 1-58415-095-5) $17.95.
Gr. 4–8. From the Unlocking the Secrets of Science series. This book describes cancer research over the years. Dr. Weinberg and his team of

researchers at the Massachusetts Institute of Technology have been at the forefront of this research. A chronology, a timeline, resources for further reading, a glossary, and an index are included.

Garcia, Kimberly. *Wilhelm Roentgen and the Discovery of X–Rays*. Bear, Del.: Mitchell Lane, 2003. 48 pp. (hc. 1-58415-114-5) $17.95.
Gr. 4–8. From the Unlocking the Secrets of Science series. While not the first scientist to produce x-rays, Roentgen was the first to understand their importance. To help assure that x-rays would be developed to the benefit of humanity he did not patent his discovery. His work won numerous awards including the Nobel Prize in Physics in 1901. A chronology, an x-ray timeline, resources for further reading, a glossary, and an index conclude the book.

Heinrichs, Ann. *Albert Einstein*. Milwaukee: World Almanac Library, 2002. 48 pp. (hc. 0-8368-5069-6) $26.60.
Gr. 4–8. From the Trailblazers of the Modern World series. The author begins by explaining why Einstein is considered by some to be "the greatest scientist of all time." Captioned photographs and sidebars, some filled with quotes from Einstein, add depth to this brief biography. A timeline, a glossary, resources for finding out more, and an index conclude the book.

Kjelle, Marylou Morano. *Raymond Damadian and the Development of MRI*. Bear, Del.: Mitchell Lane, 2003. 48 pp. (hc. 1-58415-141-2) $17.95.
Gr. 4–8. From the Unlocking the Secrets of Science series. The information on Damadian is overshadowed by the information on the development of magnetic resonance imaging (MRI) and the other scientists whose work made the MRI possible. MRI provides doctors with scans of internal organs. A Raymond Damadian chronology, a magnetic resonance imaging timeline, resources for further reading, a glossary, and an index conclude the book.

Komroff, Manuel. *Thomas Jefferson*. New Baltimore, N.Y.: Marshall Cavendish, 1991. 159 pp. (hc. 1-55905-083-7) $14.40.
Gr. 4–8. From the American Cavalcade series. This biography of America's third president, Thomas Jefferson, covers his life as a member of the Virginia House of Burgesses, a member of the Second Conti-

nental Congress, a writer of the Declaration of Independence, a governor of Virginia, and secretary of state. Especially interesting is the account of his design and construction of his home, Monticello. The book includes a list of books for further reading and an index.

Marrin, Albert. *Secrets from the Rocks: Dinosaur Hunting with Roy Chapman Andrews*. New York: Dutton Children's Books, 2002. 64 pp. (hc. 0-525-46743-2) $18.99.
Gr. 4–8. From washing floors at the American Museum of Natural History to collecting whale skeletons to uncovering fossils in the Gobi Desert, Andrews worked hard to achieve his dreams. Photographs of his expeditions and discoveries are included in the well-researched book. At the end of the book are pictures of the prehistoric animals mentioned in the book, other books to read, websites, and an index.

Mattern, Joanne. *Joseph E. Murray and the Story of the First Human Kidney Transplant*. Bear, Del.: Mitchell Lane, 2003. 48 pp. (hc. 1-58415-136-6) $17.95.
Gr. 4–8. From the Unlocking the Secrets of Science series. Murray's work led to the first successful human kidney transplant for which he received the Nobel Prize in Physiology or Medicine 1990. Murray eventually gave up kidney transplants to return to his work as a plastic surgeon. Throughout his career he traveled to foreign countries to teach other doctors his techniques and to improve the lives of their patients. The book concludes with a Joseph E. Murray chronology, a transplant timeline, resources for further reading, a glossary, and an index.

Matthews, Tom L. *Light Shining through the Mist: A Photobiography of Dian Fossey*. Washington, D.C.: National Geographic Society, 1998. 64 pp. (hc. 0-7922-7300-1) $17.95.
Gr. 4–8. Stunning color photographs and quotes from Dian Fossey's writings combine to assure that readers understand her important work in the Congo and Rwanda. Her observations and research provided a wealth of information about gorillas and her conservation efforts helped to maintain their habitats and save them from poachers. A bibliography, resources for further information, and an index conclude the book.

Pettit, Jayne. *Jane Goodall: Pioneer Researcher*. New York: Franklin Watts, 1999. 128 pp. (hc. 0-531-11522-4) $20.00.
Gr. 4–8. From A Book Report Biography series. Jane Goodall began her research on chimpanzees in the Gombe Stream Reserve at the age

of twenty-six. Since it was dangerous for her to go alone, her mother went with her. This book describes her years of study and the startling discoveries she made about chimpanzees. A chronology, notes on sources, resources for learning more, and an index conclude the book.

Riddle, John, and Jim Whiting. *Stephen Wozniak and the Story of Apple Computer*. Bear, Del.: Mitchell Lane, 2003. 48 pp. (hc. 1-58415-109-9) $17.95.
Gr. 4–8. From the Unlocking the Secrets of Science series. This brief biography chronicles the life of Steve Wozniak and the founding of Apple Computer. Readers learn that he finished his college degree years after he was a successful businessperson. Now he devotes his time to helping others and teaching children to use computers. The book concludes with a Steve Wozniak chronology, a computer time-line, resources for further reading, a glossary, and an index.

Severs, Vesta-Nadine, and Jim Whiting. *Oswald Avery and the Story of DNA*. Bear, Del.: Mitchell Lane, 2002. 48 pp. (hc. 1-58415-110-2) $17.95.
Gr. 4–8. From the Unlocking the Secrets of Science series. The authors begin with a true story that shows readers the importance of deoxyribo-nucleic acid (DNA) fingerprinting. While Avery did not discover DNA fingerprinting, his research proved that only DNA contains the genetic messages for reproduction, an important breakthrough. The book con-cludes with an Oswald Avery chronology, a DNA timeline, resources for further reading, a glossary, and an index.

Tracy, Kathleen. *William Hewlett: Pioneer of the Computer Age*. Bear, Del.: Mitchell Lane, 2003. 48 pp. (hc. 1-58415-178-1) $17.95.
Gr. 4–8. From the Unlocking the Secrets of Science series. Students who have trouble reading empathize with Hewlett's reading problems and discover how he compensated for them and went on to become one of the founders of Hewlett-Packard. A chronology, a timeline, books for further reading, a glossary, and an index conclude the book.

——. *Willem Kolff and the Invention of the Dialysis Machine*. Bear, Del.: Mitchell Lane, 2003. 48 pp. (hc. 1-58415-135-8) $17.95.
Gr. 4–8. From the Unlocking the Secrets of Science series. After watching a patient die from renal failure, Kolff began to work on a way to remove urea from the body when the kidneys fail. After creating the

artificial kidney machine, he created an artificial heart. He became known as the "Father of Artificial Organs." A Willem Kolff chronology, a dialysis and artificial heart timeline, books for further reading, a glossary, and an index conclude the book.

——. *Barbara McClintock: Pioneering Geneticist*. Bear, Del.: Mitchell Lane, 2002. 48 pp. (hc. 1-58415-111-0) $17.95.
Gr. 4–8. From the Unlocking the Secrets of Science series. At the beginning of the biography, readers meet Gregor Mendel whose research provided the basis for the study of modern genetics. It took McClintock many years before her work was recognized. In 1983, she won the Nobel Prize in Physiology or Medicine for her discovery of transposable elements in genes. The book concludes with a chronology, a timeline, resources for further reading, a glossary, and an index.

Wukovits, John. *Bill Gates: Software King*. New York: Franklin Watts, 2000. 128 pp. (hc. 0-531-11669-7) $20.00.
Gr. 4–8. From a Book Report Biography series. When Bill Gates and Paul Allen read about the first minicomputer, they wrote its creator Ed Roberts and told him they had written a programming language for the computer. When he asked for a demonstration, they spent a frantic five weeks writing the program. This engrossing account of their first success holds readers' attention as they eagerly read the rest of this informative biography. A chronology, a note on sources, resources for more information, and an index are located at the end of the book.

Zannos, Susan. *Edward Roberts and the Story of the Personal Computer*. Bear, Del.: Mitchell Lane, 2003. 48 pp. (hc. 1-58415-118-8) $17.95.
Gr. 4–8. From the Unlocking the Secrets of Science series. From working in experimental surgery to earning a degree in electrical engineering to developing a personal computer to farming to medical school, Roberts has led an interesting life that has taken many twists and turns. Whereas many students do not know who Edward Roberts is, they do know others in the computer industry that crossed his path including Steve Wozniak, Steve Jobs, Paul Allen, and Bill Gates. Included in the book are a chronology, a computer timeline, resources for further reading, a glossary, and an index.

Aldrich, Lisa J. *Cyrus McCormick and the Mechanical Reaper.* Greensboro, N.C.: Morgan Reynolds, 2002. 112 pp. (hc. 1-883846-91-9) $21.95.
Gr. 5–8. From the American Business Leaders series. McCormick's father began the design for a mechanical reaper to save time and money during the harvest. McCormick improved the design in order to get himself out of the hard work harvesting required. Readers learn of McCormick's struggles to market the reaper and how he built an empire. The book includes a glossary, websites, a selected bibliography, a timeline, and an index.

Barron, Rachel Stiffler. *Lise Meitner: Discoverer of Nuclear Fission.* Greensboro, N.C.: Morgan Reynolds, 2000. 112 pp. (hc. 1-883846-52-8) $21.95.
Gr. 5–8. From the Great Scientists series. Born in the late nineteenth century, physicist Lise Meitner faced discrimination because she was female and of Jewish decent. When her colleague Otto Hahn received the Nobel Prize for their co-discovery of nuclear fission, her contributions were ignored. This is the story of a brilliant, private person who persevered .

Byman, Jeremy. *Carl Sagan: In Contact with the Cosmos.* Greensboro, N.C.: Morgan Reynolds, 2000. 112 pp. (hc. 1-883846-55-2) $21.95.
Gr. 5–8. From the Great Scientists series. Controversial astronomer Carl Sagan made noteworthy contributions to the understanding of the solar system and engendered an interest in the solar system in the public through his teaching and television appearances. A timeline and a bibliography are included.

Fine, Edith Hope. *Barbara McClintock, Nobel Prize Geneticist.* Berkeley Heights, N.J.: Enslow, 1997. 128 pp. (hc. 0-89490-983-5) $20.95.
Gr. 5–8. From the People to Know series. Since Barbara McClintock did not have a phone in her house, the Nobel Committee could not reach her to tell her she had won the Nobel Prize in Physiology or Medicine. Instead, on the morning of October 10, 1983, she heard about her award on the radio. Since the 1940s, she had been studying the genes in maize and her research had largely been ignored. This is an informative biography about a diligent, patient woman scientist. A

chronology, chapter notes, a glossary, resources for further reading, and an index are included.

Goodhue, Thomas W. *Curious Bones: Mary Anning and the Birth of Paleontology.* Greensboro, N.C.: Morgan Reynolds, 2002. 112 pp. (hc. 1-883846-93-5) $20.95.
Gr. 5–8. Mary Anning's finds established her as an expert in the new field of paleontology at a time when few women were accepted in scientific fields. While she had little formal schooling, her father's passion for fossils lived on in her after his death and drove her to search the seashore and cliffs near her house for fossils. The book includes a glossary, a timeline, photographs of Anning's fossil finds, and a bibliography.

McPherson, Stephanie Sammartino. *Ordinary Genius: The Story of Albert Einstein.* Minneapolis: Carolrhoda Books, 2003. 96 pp. (hc. 0-87614-788-0) $25.26.
Gr. 5–8. From the Trailblazers series. McPherson depicts the very human side of Einstein, which helps readers realize that this great thinker was also a great deal like them. She also writes about the difficulties he encountered because of his Jewish ancestry. An afterword, notes, a bibliography, and an index are included.

——. *TV's Forgotten Hero: The Story of Philo Farnsworth.* Minneapolis: Carolrhoda Books, 2003. 96 pp. (hc. 0-87614-017-X) $25.26.
Gr. 5–8. As a young boy Philo was fascinated by electricity and his tinkering often left his family and neighbors in the dark. As a teenager he realized that transmitting pictures over wires could be done if the images were sent one line at a time. This well-written biography tells of his struggles to create a workable television and why he failed to receive the recognition he deserved. An afterword, notes, a bibliography, and an index conclude the book.

O'Connor, Barbara. *Leonardo da Vinci: Renaissance Genius.* Minneapolis: Carolrhoda Books, 2003. 111 pp. (hc. 0-87614-467-9) $25.26.
Gr. 5–8. Information on his most famous paintings makes fascinating reading as does the information on his scientific experiments and inventions. Reproductions, a list of locations of major works by da Vinci, quotes from him, a timeline of his life, a timeline of Renaissance Europe, very brief biographies of artists of the Italian Renaissance, re-

sources for learning more, a listing of sources, a selected bibliography, and an index are included in this well-researched biography.

Pflaum, Rosalynd. *Marie Curie and Her Daughter Irene*. Minneapolis: Lerner, 1993. 144 pp. (hc. 0-8225-4915-8) $14.95.
Gr. 5–8. Between the two of them, Marie Curie and Irene Curie won three Nobel Prizes. This biography tells the story of two women who succeeded in careers in science at a time when few women were engaged in scientific endeavors. Some of the material for the biography was obtained by interviewing the Curie family members and friends. A selected bibliography and an index are included.

Pflueger, Lynda. *George Eastman: Bringing Photography to the People*. Berkeley Heights, N.J.: Enslow, 2002. 128 pp. (hc. 0-7660-1617-X) $20.95.
Gr. 5–8. From the Historical American Biographies series. Eastman did not invent the camera, but he did contribute a number of inventions that made cameras easier to use. Sidebars interspersed throughout the text contain a wide array of intriguing tidbits about Eastman's life and work. A chronology, chapter notes, books for further reading, Internet addresses, and an index are included.

Warrick, Karen Clemens. *John Muir: Crusader for the Wilderness*. Berkeley Heights, N.J.: Enslow, 2002. 128 pp. (hc. 0-7660-1622-6) $20.95.
Gr. 5–8. From the Historical American Biographies series. Muir is remembered for his deep respect and appreciation for nature, for being a founding member of the Sierra Club, and for discovering a glacier in the Sierra Nevada Mountains. As a child, he worked very hard on the family farm and invented several items including a device to get him out of bed early in the morning so that he could read and study. Black and white photographs, sidebars, a chronology, chapter notes, books for further reading, Internet addresses, and an index are provided.

Anderson, Margaret J. *Charles Darwin: Naturalist*. Berkeley Heights, N.J.: Enslow, 1994. 128 pp. (hc. 0-89490-476-0) $20.95; (pbk. 0-7660-1868-7) $10.95.
Gr. 5–12. From the Great Minds of Science series. One of the most important scientific ideas to come out of the nineteenth century was the theory of evolution. The quiet naturalist who sparked a storm with this

new theory and opened up a completely new way of thinking for scientists was Charles Darwin. His book, *The Origin of Species*, describes his theory of evolution, which extends his grandfather's theory of evolution. Darwin's years of sailing and exploring on *H.M.S. Beagle* and his years of experimenting and studying helped him develop and refine his theory. This book concludes with a chronology, chapter notes, a glossary, books for further reading, Internet addresses, and an index.

——. *Carl Linnaeus: Father of Classification.* Berkeley Heights, N.J.: Enslow, 1997. 128 pp. (hc. 0-89490-786-7) $20.95.
Gr. 5–12. From the Great Minds of Science series. Carl Linnaeus developed a system for naming living things that helped scientists describe every living thing. This system, still used today, is known as binominal nomenclature. Carl's parents wanted him to become a pastor like his father and grandfather, but his love from early childhood had been plants. Carl was also an explorer and once spent four months exploring Lapland. The book ends with a chronology, chapter notes, a glossary, books for further reading, Internet addresses, and an index.

——. *Isaac Newton: The Greatest Scientist of All Time.* Berkeley Heights, N.J.: Enslow, 1996. 128 pp. (hc. 0-89490-681-X) $20.95.
Gr. 5–12. From the Great Minds of Science series. Isaac Newton was an English scientist, astronomer, and mathematician. He is considered by many to be the greatest scientist of all time. There were many things in the universe that fascinated Newton—especially light and motion. Using a prism, Newton first discovered that sunlight is made up of light rays of many different colors. Today, scientists work out complex ideas using calculus, the branch of mathematics developed by Newton. The book ends with a chronology, chapter notes, a glossary, books for further reading, Internet addresses, and an index.

Andronik, Catherine M. *Copernicus: Founder of Modern Astronomy.* Berkeley Heights, N.J.: Enslow, 2002. 128 pp. (hc. 0-7660-1755-9) $20.95.
Gr. 5–12. From the Great Minds of Science series. Centuries ago people believed that the earth stood still and the sun revolved around it. Because of the work of Copernicus the idea of the earth circling the sun began to be universally accepted. Primitive scientific instruments were used to prove this concept before the invention of the telescope. His work in astronomy and physics made him one of the great minds of

science. The book concludes with a chronology, chapter notes, a glossary, books for further reading, Internet addresses, and an index.

Datnow, Claire. *Edwin Hubble: Discoverer of Galaxies.* Berkeley Heights, N.J.: Enslow, 1997. 128 pp. (hc. 0-89490-934-7) $20.95; (pbk. 0-76601-864-4) $10.95.
Gr. 5–12. From the Great Minds of Science series. Students familiar with the magnificent photographs beamed to Earth by the Hubble Space Telescope will discover in this biography why the telescope was named after this noted astronomer. One of America's greatest astronomers, Edwin Hubble discovered that the Milky Way was one of many galaxies in the universe. His discovery showed scientists that the universe was much larger than had ever before been imagined. An epilogue, activities, a chronology, chapter notes, a glossary, books for further reading, and an index are included.

Dolan, Ellen M. *Thomas Alva Edison: Inventor.* Berkeley Heights, N.J.: Enslow, 1998. 128 pp. (hc. 0-7660-1014-7) $20.95.
Gr. 5–12. From the Historical American Biographies series. While well known for inventing the electric light bulb and the phonograph, many readers are not aware of his extensive background in telegraphy. A close friend of Henry Ford and Harvey Firestone, Edison liked camping out West each year with them to share ideas. The book ends with a chronology, chapter notes, a glossary, books for further reading, Internet sites, and an index.

Flammang, James M. *Robert Fulton: Inventor and Steamboat Builder.* Berkeley Heights, N.J.: Enslow, 1999. 128 pp. (hc. 0-7660-1141-0) $20.95.
Gr. 5–12. From the Historical American Biographies series. As a boy, Robert Fulton is believed to have invented a rocket used in a town celebration, lead pencils, and household utensils. He was also good at "gunsmithing" and manufactured an air-gun. Fulton was enchanted with art and became quite accomplished in drawing and portraiture. He is remembered for making improvements to the steam engine that made operating a steamboat a profitable venture. The book ends with a chronology, chapter notes, a glossary, books for further reading, Internet sites, and an index.

Foster, Leila Merrell. *Benjamin Franklin: Founding Father and Inventor*. Berkeley Heights, N.J.: Enslow, 1997. 128 pp. (hc. 0-89490-784-0) $20.95.

Gr. 5–12. From the Historical American Biographies series. Much is known about Benjamin Franklin's middle and later years, but this book gives information about his childhood and his numerous interests throughout his life. Because the father's estate was always left to the oldest son, and because Benjamin was the youngest son, he searched diligently for ways to earn a good living. The book ends with a chronology, chapter notes, a glossary, books for further reading, and an index.

Goldenstern, Joyce. *Albert Einstein: Physicist and Genius*. Berkeley Heights, N.J.: Enslow, 1997. 128 pp. (hc. 0-89490-480-9) $20.95; (pbk. 0-7660-1869-5) $10.95.

Gr. 5–12. From the Great Minds of Science series. As students read about the life and work of this noted scientist, they also learn about how he built on the work of others including Galileo and Planck. When the Nazis came to power in Germany, Einstein was forced to flee for his life. Because of his reputation as a well-known scientist, he was welcomed at the Institute for Advanced Study in Princeton, New Jersey. This biography contains experiments and activities to help students comprehend Einstein's theories. Included in the book are a chronology, notes, a glossary, books for further reading, and an index.

Gow, Mary. *Tycho Brahe: Astronomer*. Berkeley Heights, N.J.: Enslow, 2002. 128 pp. (hc. 0-7660-1757-5) $20.95.

Gr. 5–12. From the Great Minds of Science series. In 1576, King Frederick II of Denmark gave the island of Hven to Tycho Brahe who built the world's first modern observatory there. He measured stars, observed comets, and recorded the first observation of a supernova. His observations were the most accurate ones in the world at his time. Among the things included in this book are a chronology, chapter notes, a glossary, books for further reading, Internet addresses, and an index.

Hamilton, Janet. *Lise Meitner: Pioneer of Nuclear Fission*. Berkeley Heights, N.J.: Enslow, 2002. 128 pp. (hc. 0-7660-1756-7) $20.95.

Gr. 5–12. From the Great Minds of Science series. Lise Meitner was the first scientist to realize that an atom could be split—a process known as nuclear fission. Prior to that time, scientists had assumed that when they bombarded uranium with neutrons the particles would add to the element's mass. Nevertheless, this is not what happened. Meitner's discovery changed all that. In spite of the difficulty during this time that a woman had in getting educated, Lise became one of the greatest scientists of all time. The book includes activities, a chronology, chapter notes, a glossary, books for further reading, Internet addresses, and an index.

Hightower, Paul. *Galileo: Astronomer and Physicist.* Berkeley Heights, N.J.: Enslow, 1997. 128 pp. (hc. 0-89490-787-5) $20.95.
Gr. 5–12. From the Great Minds of Science series. Called by many the "founder of modern experimental science," Galileo is one of the greatest scientists of all time. He discovered the law of the pendulum and the law of freely falling bodies, proving that the earth and the planets revolve around the sun. He also built a telescope and discovered four of Jupiter's moons. Using a chronology, chapter notes, a glossary, books for further reading, Internet addresses, and an index, this book examines the life and work of this brilliant scientist.

Klare, Roger. *Gregor Mendel: Father of Genetics.* Berkeley Heights, N.J.: Enslow, 1997. 128 pp. (hc. 0-89490-789-1) $20.95.
Gr. 5–12. From the Great Minds of Science series. Gregor Mendel was an Austrian monk who became the father of genetics, the science of explaining how parents pass certain characteristics on to their offspring. Mendel's work began when he sought answers to his questions about plant hybrids, but years later others would study his work to learn about heredity. Activities, a chronology, chapter notes, a glossary, resources for further reading, and an index are included.

McCarthy, Pat. *Henry Ford: Building Cars for Everyone.* Berkeley Heights, N.J.: Enslow, 2002. 128 pp. (hc. 0-7660-1620-X) $20.95.
Gr. 5–12. From the Historical American Biographies series. In 1903 the Ford Motor Company was organized by Henry Ford. At the age of forty and after countless inventions and trials, he manufactured cars that the average man could afford. He had many friends (the most notable of whom was Thomas Alva Edison) and he was respected by almost everyone. He was a pacifist who spoke ardently against World War I. The book includes sidebars of additional information, a chronology, chapter notes, a glossary, books for further reading, Internet addresses, and an index.

Old, Wendie C. *The Wright Brothers: Inventors of the Airplane*. Berkeley Heights, N.J.: Enslow, 2000. 128 pp. (hc. 0-7660-1095-3) $20.95.
Gr. 5–12. From the Historical American Biographies series. Perhaps one of the greatest influences on the Wright brothers was their family. They were always immersed in their sense of family and were always supported in everything they did by their family. Few people know that their first successful endeavor was in printing. They next branched into bicycles, which led them to their fascination with manned flight. There are also sidebars of information in the book. Concluding this volume are a chronology, chapter notes, a glossary, books for further reading, Internet addresses, and an index.

Pasachoff, Naomi. *Niels Bohr: Physicist and Humanitarian*. Berkeley Heights, N.J.: Enslow, 1997. 128 pp. (hc. 0-89490-788-3) $20.95.
Gr. 5–12. From the Great Minds of Science series. One of the first scientists to explain atomic structure accurately and to add a great deal to the development of quantum theory was Niels Bohr. His work eventually helped lead to many devices like lasers and transistors. Bohr was also a strong supporter of open communications and cooperation between all the different countries of the world. This book explores his life and his work. It concludes with a chronology, chapter notes, a glossary, books for further reading, Internet addresses, and an index.

——. *Marie Curie and the Science of Radioactivity*. New York: Oxford University Press, 1996. 109 pp. (hc. 0-19-509214-7) $24.00; (pbk. 0-19-512011-6) $11.95.
Gr. 5–12. From the Oxford Portraits in Science series. Curie's years of work with radioactive substances earned her two Nobel Prizes and claimed her life. This is the biography of a quiet, unassuming woman. Black and white photographs, a chronology, books for further reading, and an index are included.

Poynter, Margaret. *Marie Curie: Discoverer of Radium*. Berkeley Heights, N.J.: Enslow, 1994. 128 pp. (hc. 0-89490-477-9) $20.95.
Gr. 5–12. From the Great Minds of Science series. One of the most famous women of all time is Marie Curie. She was born in Poland at a time where it was most unusual for a Polish person, especially a woman, to obtain an education, so she moved to France. She was one of the first European women to receive a doctorate degree from the Sorbonne. She married Pierre Curie and they became fascinated with

the phenomenon of radiation. Marie and Pierre won the Nobel Prize in Physics in 1903 and Marie won the Nobel Prize in Chemistry in 1911. Included in this book are a chronology, chapter notes, a glossary, books for further reading, Internet addresses, and an index.

——. *The Leakeys: Uncovering the Origins of Humankind.* Berkeley Heights, N.J.: Enslow, 1997. 128 pp. (hc. 0-89490-788-3) $20.95; (pbk. 0-76601-873-3) $10.95.
Gr. 5–12. From the Great Minds of Science series. Louis, Mary, Richard, and Maeve Leakey are the most famous names in the field of paleoanthropology, the study of the earliest ancestors of humankind. They believed that early humans used simple tools and walked upright like earlier apelike creatures. Their work and discoveries about human evolution has not been surpassed. Included in this book are a chronology, chapter notes, a glossary, books for further reading, Internet addresses, and an index.

Schuman, Michael A. *Alexander Graham Bell: Inventor and Teacher.* Berkeley Heights, N.J.: Enslow, 1999. 128 pp. (hc. 0-7660-1096-1) $20.95.
Gr. 5–12. From the Historical American Biographies series. Alexander Graham Bell was a multitalented inventor who preferred to be remembered as a teacher of the deaf. An accidental acid spill in the laboratory caused Bell to summon his assistant in the next room. When Watson came running, they realized that he heard Bell's voice on the telephone! In addition to sidebars of information, the book also has a chronology, chapter notes, places to visit, a glossary, books for further reading, Internet addresses, and an index.

Shapiro, Miles J. *Charles Drew: Life-Saving Scientist.* Austin, Tex.: Raintree Steck-Vaughn, 1997. 112 pp. (hc. 0-8172-4403-4) $20.98.
Gr. 5–12. From the Innovative Minds series. Charles Drew is remembered not only for his groundbreaking work in preserving and storing blood, but also for his work as a teacher at Howard University. He was determined that African American doctors would have the best possible education. A glossary, books for further reading, sources, and an index are included.

Smith, Linda Wasmer. *Louis Pasteur: Disease Fighter.* Berkeley Heights, N.J.: Enslow, 1997. 128 pp. (hc. 0-89490-790-5) $20.95.

Gr. 5–12. From the Great Minds of Science series. Some microorganisms (called germs by many) are very helpful to humankind; some cause grapes to ferment and become wine and others cause milk to ripen and become cheese. Louis Pasteur was one of the first scientists to understand this. He developed what became known as pasteurization, which destroyed harmful bacteria in liquids such as milk and wine. The book concludes with a chronology, chapter notes, a glossary, books for further reading, Internet addresses, and an index.

Streissguth, Thomas. *Rocket Man: The Story of Robert Goddard.* Minneapolis: Carolrhoda Books, 2003. 88 pp. (hc. 0-87614-863-1) $25.26.
Gr. 5–12. While his achievements were not fully appreciated in his lifetime, Goddard's liquid fuel rocket led the way for the development of satellites, spacecraft, and guided missiles. Black and white photographs accompany the interesting, informative text. An afterword, notes, a glossary, a bibliography, and an index conclude the book.

Tocci, Salvatore. *Alexander Fleming: The Man Who Discovered Penicillin.* Berkeley Heights, N.J.: Enslow, 2002. 128 pp. (hc. 0-7660-1998-5) $30.00.
Gr. 5–12. From the Great Minds of Science series. The compelling introduction to this biography tells how penicillin saved the lives of people who were dying from infections that ravaged their bodies. As Tocci tells the story of Fleming's work, he helps readers understand the dramatic difference Fleming's discoveries had on the practice of medicine. At the end of the story, students read of the problems caused by the overuse of antibiotics such as penicillin. Activities, a chronology, chapter notes, a glossary, resources for learning more, and an index conclude the book.

——. *Jonas Salk: Creator of the Polio Vaccine.* Berkeley Heights, N.J.: Enslow, 2003. 128 pp. (hc. 0-7660-2097-5) $20.95.
Gr. 5–12. From the Great Minds of Science series. In the early to mid-1900s, the United States suffered many epidemics of the dreaded disease polio. It caused paralysis or death and children were the most likely ones to contract this disease. Fear was rampant across the U.S. However, in 1955 Salk created a polio vaccine. This book is distinguished by the inclusion of a chronology, chapter notes, a glossary, books for further reading, Internet addresses, and an index.

Yannuzzi, Della A. *Madam C. J. Walker: Self-Made Businesswoman.* Berkeley Heights, N.J.: Enslow, 2000. 112 pp. (hc. 0-7660-1204-2) $18.95.

Gr. 5–12. From the Great Minds of Science series. Born into poverty this amazing woman invented a hair care product for African Americans and turned the product into a successful cosmetic business. She helped other women become successful by hiring them to sell her products and she became a philanthropist. Located at the end of the book are a chronology, chapter notes, resources for further reading, and an index.

Yount, Lisa. *William Harvey: Discoverer of How Blood Circulates.* Berkeley Heights, N.J.: Enslow, 1994. 128 pp. (hc. 0-89490-481-7) $20.95.

Gr. 5–12. From the Great Minds of Science series. William Harvey showed the world that the heart was actually a pump, and that blood circulates throughout the body. This was one of the great advances in the study of medicine. Harvey learned many things by dissecting the bodies of people who had died. This book includes an afterword, activities, a chronology, chapter notes, a glossary, books for further reading, Internet addresses, and an index.

——. *Antoine Lavoisier: Founder of Modern Chemistry.* Berkeley Heights, N.J.: Enslow, 1997. 128 pp. (hc. 0-089490-785-9) $20.95.

Gr. 5–12. From the Great Minds of Science series. Before the French Revolution, there was a wealthy and important man in France known as Antoine Lavoisier. Throughout the Western world, his experiments on combustion were well known. One of the people he shared his ideas with was Benjamin Franklin. With the aid of three other chemists, Lavoisier devised a system for naming chemicals. Much of his work was recorded in his book *Elements of Chemistry.* An afterword, activities, a chronology, chapter notes, a glossary, books for further reading, and an index conclude the book.

——. *Antoni van Leeuwenhoek: First to See Microscopic Life.* Berkeley Heights, N.J.: Enslow, 1996. 128 pp. (hc. 0-089490-680-1) $20.95.

Gr. 5–12. From the Great Minds of Science series. In the introduction students learn that Antoni van Leeuwenhoek did not invent the microscope, but with it peered into an exciting new world, a world that no one knew existed. He was the first to see bacteria, red blood cells, and more kinds of invisible "animals" than anyone knew existed. He is considered one of the great minds of science because of his discovery that tiny microbes cause disease. Black and white photographs and drawings enable readers to appreciate the tiny organisms Leeuwenhoek saw under his microscope. An afterword, activities, a chronology, chapter notes, a glossary, books for further reading, and an index conclude the book.

Allan, Tony. *Isaac Newton.* Chicago: Heinemann Library, 2001. 48 pp. (hc. 1-58810-053-7) $27.00.
Gr. 6–8. From the Groundbreakers series. Newton's students considered him a poor teacher and many did not attend his lectures. Information such as this grabs readers' attention as they learn about Newton and his discoveries. Photographs, reproductions, diagrams, sidebars, a timeline, more books to read, a glossary, and an index complement the information presented in the text.

Champion, Neil. *Charles Babbage.* Chicago: Heinemann Library, 2000. 48 pp. (hc. 1-57572-367-0) $27.00.
Gr. 6–8. From the Groundbreakers series. From Babbage's Analytical Engine came the idea for the computer. This compact biography contains an overview of his life and his work. Photographs, reproductions, diagrams, and sidebars enhance the text. A timeline, more books to read, a glossary, and an index conclude the book.

——. *James Watt.* Chicago: Heinemann Library, 2000. 48 pp. (hc. 1-57572-371-9) $27.00.
Gr. 6–8. From the Groundbreakers series. Readers discover in this biography that Watt did not invent the steam engine, but his improvements did make it commercially viable. Photographs, reproductions, diagrams, and sidebars enhance the text. A timeline, more books to read, a glossary, and an index conclude the book.

Fullick, Ann. *Marie Curie.* Chicago: Heinemann Library, 2000. 48 pp. (hc. 1-57572-374-23) $27.00.
Gr. 6–8. From the Groundbreakers series. In this compact biography students learn of Curie's tireless work on radiation and her two Nobel Prizes. Photographs, reproductions, diagrams, and sidebars enhance the text. A timeline, more books to read, a glossary, and an index conclude the book.

——. *Charles Darwin*. Chicago: Heinemann Library, 2000. 48 pp. (hc. 1-57572-368-9) $27.00.
Gr. 6–8. From the Groundbreakers series. Focusing on Darwin's five-year voyage aboard *H.M.S. Beagle* helps readers understand how he gathered the research needed to support his theory of evolution. Photographs, reproductions, diagrams, and sidebars enhance the text. A timeline, more books to read, a glossary, and an index conclude the book.

——. *Michael Faraday*. Chicago: Heinemann Library, 2000. 48 pp. (hc. 1-57572-375-1) $27.00.
Gr. 6–8. From the Groundbreakers series. Faraday's work in electricity and magnetism are examined in this concise biography. Photographs, reproductions, diagrams, and sidebars enhance the text. A timeline, more books to read, a glossary, and an index conclude the book.

——. *Louis Pasteur*. Chicago: Heinemann Library, 2000. 48 pp. (hc. 1-57572-373-5) $27.00.
Gr. 6–8. From the Groundbreakers series. Students discover the far-reaching impact of Pasteur's research on medicine and science as they read about his life. Photographs, reproductions, diagrams, and sidebars enhance the text. A timeline, more books to read, a glossary, and an index conclude the book.

Macdonald, Fiona. *Edwin Hubble*. Chicago: Heinemann Library, 2001. 48 pp. (hc. 1-58810-055-2) $27.00.
Gr. 6–8. From the Groundbreakers series. Hubble did not get along well with some other astronomers, but they respected his research on things such as the age of the universe. Photographs, reproductions, diagrams, sidebars, a timeline, more books to read, a glossary, and an index complement the information presented in the text.

Malam, John. *Florence Nightingale*. Chicago: Heinemann Library, 2001. 48 pp. (hc. 1-58810-051-0) $27.00.
Gr. 6–8. From the Groundbreakers series. Nightingale's pioneering work toward establishing nursing as a profession for women is highlighted in this brief biography. Photographs, reproductions, diagrams, sidebars, a timeline, more books to read, a glossary, and an index complement the information presented in the text.

Mason, Paul. *Galileo*. Chicago: Heinemann Library, 2001. 48 pp. (hc.
    1-58810-052-9) $27.00.
Gr. 6–8. From the Groundbreakers series. Galileo's illnesses and his
conflicts with the Catholic Church are described as well as his studies
of the heavens. Photographs, reproductions, diagrams, sidebars, a timeline, more books to read, a glossary, and an index complement the information presented in the text.

Parker, Steve. *Alexander Fleming*. Chicago: Heinemann Library, 2001.
    48 pp. (hc. 1-58810-050-2) $27.00.
Gr. 6–8. From the Groundbreakers series. Parker describes Fleming's
work: most notably his discovery of the beneficial uses of penicillin
and his promotion of the use of penicillin. Photographs, reproductions,
diagrams, sidebars, a timeline, more books to read, a glossary, and an
index complement the information presented in the text.

Reid, Struan. *John Logie Baird*. Chicago: Heinemann Library, 2000. 48
    pp. (hc. 1-57572-372-7) $27.00.
Gr. 6–8. From the Groundbreakers series. The focus of this book is on
Baird's inventions, primarily the development of television. Farnsworth's work on television in America is not mentioned. Photographs,
reproductions, diagrams, and sidebars enhance the text. A timeline,
more books to read, a glossary, and an index conclude the book.

——. *Alexander Graham Bell*. Chicago: Heinemann Library, 2001. 48
    pp. (hc. 1-57572-366-2) $27.00.
Gr. 6–8. From the Groundbreakers series. After inventing the telephone
and establishing the Bell Telephone Company, Bell resigned in order to
pursue his real interest, inventing. Readers also learn about his work
with the deaf. Photographs, reproductions, diagrams, sidebars, a timeline, more books to read, a glossary, and an index complement the information presented in the text.

——. *Albert Einstein*. Chicago: Heinemann Library, 2000. 48 pp. (hc. 1-
    57572-365-4) $27.00.
Gr. 6–8. From the Groundbreakers series. Readers discover how Einstein continuously questioned and searched for answers about the universe. His work is explained in easy-to-understand terms. The book
includes photographs, reproductions, diagrams, sidebars, a timeline,
more books to read, a glossary, and an index.

Tagliaferro, Linda. *Thomas Edison: Inventor in the Age of Electricity.*
Minneapolis: Lerner, 2003. 128 pp. (hc. 0-8225-4689-2) $25.26.
Gr. 6–8. From the Lerner Biography series. Twenty-hour workdays,
perseverance, and natural genius enabled Edison to blaze the way to the
commercial use of electricity and changes in the entertainment industry
with his phonograph and motion picture camera. Included in the book
are black and white photographs, a timeline, an extensive list of resources for learning more, and an index.

Williams, Brian. *Thomas Alva Edison.* Chicago: Heinemann Library,
2000. 48 pp. (hc. 1-57572-377-8) $27.00.
Gr. 6–8. From the Groundbreakers series. Williams describes Edison's
invention of the phonograph, electric lighting, and motion pictures. All
things we take for granted today. Photographs, reproductions, diagrams, sidebars, a timeline, more books to read, a glossary, and an index are included.

Birch, Beverly. *Alexander Fleming: Pioneer with Antibiotics.* Woodbridge, Conn.: Blackbirch Press, 2002. 64 pp. (hc. 1-56711-656-6)
$27.44.
Gr. 6–10. From the Giants of Science series. Photographs, drawings,
and quotes in sidebars enhance the story of this scientist whose messy
habits led to the discovery of the potential of penicillin to kill bacteria.
The book includes a timeline, resources for learning more, a glossary,
and an index.

——. *Guglielmo Marconi: Radio Pioneer.* Woodbridge, Conn.: Blackbirch Press, 2001. 64 pp. (hc. 1-56711-337-60) $27.44.
Gr. 6–10. From the Giants of Science series. Marconi's mind seemed to
always be filled with questions that required answers. When in the hospital with a sore throat he and a friend used the long corridors for experiments. His thirty-six-foot boat became a floating laboratory. His
questions and his experiments enabled him to achieve his dream of
worldwide communications. A timeline, a glossary, resources for more
information, and an index are included.

Boerst, William J. *Tycho Brahe: Mapping the Heavens.* Greensboro,
N.C.: Morgan Reynolds, 2003. 128 pp. (hc. 1-883846-97-8)
$23.95.

Gr. 6–10. From the Renaissance Scientists series. During the sixteenth century the Danish astronomer Tycho Brahe studied the planets and stars and left extensive written records of his findings. Color illustrations, a timeline, a list of sources, a list of books, and websites are included.

Burgan, Michael. *Henry Ford*. Chicago: Ferguson Publishing, 2001. 127 pp. (hc. 0-89434-369-6) $16.95.
Gr. 6–10. From the Ferguson Career Biography series. Ford's contributions to the automobile industry included making better cars and making affordable cars. This biography explains how he did that and provides an intriguing look into the development of the Ford Motor Company. The end of the book contains a variety of resources including a timeline of Henry Ford's life, how to become an entrepreneur, resources for learning more about entrepreneurs, how to become a business manager, resources for learning more about business managers, resources for learning more about Henry Ford, and an index.

Conley, Kevin. *Benjamin Banneker: Scientist and Mathematician*. New York: Chelsea House, 1989. 124 pp. (hc. 1-55546-573-0) $23.95; (pbk. 0-7910-0231-4) $9.95.
Gr. 6–10. From the Black Americans of Achievement series. A self-taught astrologer, Banneker wrote almanacs and this biography begins by describing the importance of almanacs to help readers realize what an accomplishment this was. Pages from his almanac, photographs, diagrams, and reproductions are placed throughout the book. Readers learn about the challenges he faced as a free Black in the time of slavery and they develop an understanding of the kind, intelligent nature of this scientist and mathematician. A chronology, books for further reading, and an index conclude the book.

Macdonald, Fiona. *Albert Einstein: Genius behind the Theory of Relativity*. Woodbridge, Conn.: Blackbirch Press, 2000. 64 pp. (hc. 1-56711-330-3) $27.44.
Gr. 6–10. From the Giants of Science series. Short entries, photographs, and reproductions provide clear, succinct descriptions of Einstein's early years, his numerous patents, and his scientific theories that changed the way we see our world. A chronology, resources for learning more, a glossary, and an index conclude the book.

Pollard, Michael. *Alexander Graham Bell: Father of Modern Communication*. Woodbridge, Conn.: Blackbirch Press, 2002. 64 pp. (hc. 1-56711-334-6) $27.44.
Gr. 6–10. From the Giants of Science series. The biography begins with the gripping story of the race to secure the first patent for the telephone, which Bell received at the age of twenty-nine. Once readers are hooked this chronicle of Bell's life and work does not disappoint them. Photographs, reproductions, and quotes in sidebars intensify the impact of the text. A chronology, resources for learning more, a glossary, and an index conclude the book.

Sproule, Anna. *Charles Darwin: Visionary Behind the Theory of Evolution*. Woodbridge, Conn.: Blackbirch Press, 2002. 64 pp. (hc. 1-56711-655-8) $27.44.
Gr. 6–10. From the Giants of Science series. Brief chunks of information each with their own subheading tell the story of Darwin's work on the theory of evolution and describe his personal life. Photographs, drawings, reproductions, and maps illustrate the text. A timeline, a glossary, resources for more information, and an index conclude the book.

——. *Thomas Edison: The World's Greatest Inventor*. Woodbridge, Conn.: Blackbirch Press, 2000. 64 pp. (hc. 1-56711-331-1) $27.44.
Gr. 6–10. From the Giants of Science series. Readers may be surprised to learn that this prolific inventor had hearing problems. Sproule places Edison's life and inventions in the context of history and shows readers how his inventions changed society. A chronology, resources for learning more, a glossary, and an index conclude the book.

——. *James Watt: Master of the Steam Engine*. Woodbridge, Conn.: Blackbirch Press, 2001. 64 pp. (hc. 1-56711-338-9) $27.44.
Gr. 6–10. From the Giants of Science series. Watt's steam engines harnessed the power of steam and helped to usher in the Industrial Revolution. Quotes, diagrams, photographs, and reproductions enhance readers' understanding of Watt's life and work. A chronology, resources for learning more, a glossary, and an index conclude the book.

Yount, Lisa. *Louis Pasteur*. San Diego: Lucent, 1994. 95 pp. (hc. 1-56006-051-4) $27.45.

Gr. 6–10. From The Importance Of Series. While probably most remembered for the development of pasteurization, Pasteur is also remembered for the development of the rabies vaccine. Other research interests included silkworm diseases, anthrax, and fermentation. Yount presents a well-rounded portrait of a man devoted to his work. A chronology of Pasteur's life, notes, books for further reading, a bibliography, and an index are included.

Ayer, Eleanor. *Lewis Latimer: Creating Bright Ideas*. Austin, Tex.: Raintree Steck-Vaughn, 1997. 112 pp. (hc. 0-8172-4407-7) $27.00.
Gr. 6–12. From the Innovative Minds series. Working as a draftsman for a firm of patent lawyers stimulated Latimer's creativity and he began improving on other people's inventions. He also worked for the United States Electric Lighting Company and for the Edison Electric Light Company. While working at these companies, he was encouraged to experiment and invent. Photographs, reproductions, a glossary, books for further reading, a list of sources, and an index are included.

Christianson, Gale E. *Isaac Newton and the Scientific Revolution*. New York: Oxford University Press, 1996. 155 pp. (hc. 0-19-509224-4) $24.00.
Gr. 6–12. From the Oxford Portraits in Science series. Newton's father died before he was born and when his mother remarried, she abandoned Newton to the care of his maternal grandmother, and so begins the story of this noted scientist. Throughout this biography, readers learn about Newton's temperamental personality as well as his work and his discoveries. Black and white photographs and reproductions are included. A chronology, books for further reading, and an index conclude the book.

Collier, Bruce, and James MacLachlan.. *Babbage and the Engines of Perfection*. New York: Oxford University Press, 1998. 123 pp. (hc. 0-19-508997-9) $28.00; (pbk. 0-19-514287-X) $11.95.
Gr. 6–12. From the Oxford Portraits in Science series. In June 1822, Charles Babbage created a small model of a calculating machine. He began working on constructing a full-scale machine, but the cost was prohibitive and so the machine was abandoned. Babbage's Analytical Engine inspired the idea for the computer. Black and white photographs, a chronology, books for further reading, and an index are included.

Dash, Joan. *The Longitude Prize*. New York: Farrar, Straus and Giroux, 2000. 200 pp. (hc. 0-374-34636-4) $16.00.
Gr. 6–12. John Harrison, carpenter and clock maker, eventually won the prize for discovering an accurate, reliable way to measure longitude. It took forty years to overcome obstacles he encountered due in part to his own personality and to the snobbishness of the British aristocracy. Clever pencil sketches illustrate the text, which ends with a timeline, a glossary, and an index.

Dommermuth-Costa, Carol. *Nikola Tesla: A Spark of Genius*. Minneapolis: Lerner, 2003. 144 pp. (hc. 0-8225-4920-4) $25.26.
Gr. 6–12. Tesla's greatest contribution was the development of alternating current that makes it possible for us to turn on lights and listen to the radio. Born in Croatia, Tesla eventually moved to America and for a brief time worked with Thomas Edison, who was to become an archrival. Black and white photographs and technical drawing accompany the story of this complex electrical engineer. A bibliography and an index conclude the book.

Edelson, Edward. *Francis Crick and James Watson and the Building Blocks of Life*. New York: Oxford University Press, 2000. 110 pp. (hc. 0-19-5115451-5) $28.00; (pbk. 0-19-513971-2) $11.95.
Gr. 6–12. From the Oxford Portraits in Science series. Crick and Watson described the structure of DNA and the impact of that discovery is still being explored by scientists around the world. One of the sidebars in the book describes cloning experiments. Black and white photographs, a chronology, books for further reading, and an index are included.

Fleming, Candace. *Ben Franklin's Almanac: Being a True Account of the Good Gentleman's Life*. New York: Simon & Schuster, 2003. 128 pp. (hc. 0-689-83549-3) $19.95.
Gr. 6–12. The format of the book is modeled after Franklin's almanac and includes quotes, etchings, and illustrations of artifacts. The unique format of the book encourages readers to explore its pages and learn about Benjamin Franklin.

Hager, Tom. *Linus Pauling and the Chemistry of Life*. New York: Oxford University Press, 1998. 142 pp. (hc. 0-19-518053-1) $28.00; (pbk. 0-19-513972-0) $11.95.

Gr. 6–12. From the Oxford Portraits in Science series. Pauling's first Nobel Prize was for his lifetime of work in chemistry and his second Nobel Prize was for his work on banning atmospheric tests of the atom bomb. The book includes text boxes that help readers understand Pauling's work. Black and white photographs, a chronology, books for further reading, and an index are included.

Jezer, Marty. *Rachel Carson.* New York: Chelsea House, 2003. 112 pp. (hc. 1-55546-646-X) $22.95; (pbk. 0-7910-0413-9) $17.00.
Gr. 6–12. From the American Women of Achievement series. The book begins with the controversy caused by the publication of Carson's *Silent Spring.* This book documented the harmful effects of pesticides on the environment. When her research and writing could not be disputed on factual grounds, she was subjected to personal ridicule. This is an interesting biography of a courageous woman. Books for further reading, a chronology, and an index are provided.

Kronstadt, Janet. *Florence Sabin, Medical Researcher.* New York: Chelsea House, 1990. 112 pp. (pbk. 0-7910-0450-3) $17.00.
Gr. 6–12. From the American Women of Achievement series. She is remembered for her work in lymphatics and tuberculosis. Throughout her career Sabin faced gender discrimination, for example a department head position was given to one of her former students who was male even though she was more qualified. A chronological list of publications, books for further reading, a chronology, and an index are included.

Markham, Lois. *Jacques-Yves Cousteau: Exploring the Wonders of the Deep.* Austin, Tex.: Raintree Steck-Vaughn, 1997. 112 pp. (hc. 0-8172-4404-2) $27.00.
Gr. 6–12. From the Innovative Minds series. Cousteau's diving inventions and his explorations made the world aware of the beauty and the fragility of the sea. Details about some of his inventions and his work make interesting reading and encourage students to further explore the life of this French oceanographer. Photographs, a glossary, books for further reading, a bibliography, and an index are included.

Muckenhoupt, Margaret. *Sigmund Freud: Explorer of the Unconscious.* New York: Oxford University Press, 1997. 157 pp. (pbk. 0-19-513212-2) $11.95.

Gr. 6–12. From the Oxford Portraits in Science series. Clear, succinct writing describes Freud's life, work, and the era during which he lived. His groundbreaking work in psychoanalysis is described, as are some of his actual cases and methods. Black and white photographs, sidebars, a chronology, books for further reading, and an index are included.

Ravage, Barbara. *George Westinghouse: A Genius for Invention.* Austin, Tex.: Raintree Steck-Vaughn, 1997. 112 pp. (hc. 0-8172-4402-6) $27.00.
Gr. 6–12. From the Innovative Minds series. Westinghouse was not only a successful inventor, he was also a successful businessman. He obtained three hundred and sixty-one patents including ones for a rotary steam engine, railroad train brakes, and electric power. Diagrams of his inventions, text boxes of additional information, photographs, and quotes from primary sources provide a resource-filled biography. A glossary, books for further reading, sources, and an index conclude the book.

Severance, John B. *Einstein: Visionary Scientist.* New York: Clarion Books, 1999. 144 pp. (hc. 0-395-93100-2) $15.00.
Gr. 6–12. The author reveals that Einstein was nominated six times for the Nobel Prize for his theory of relativity, but was not awarded the prize because the scientists on the award committee could not understand his calculations. Eventually, he was awarded the Nobel Prize "for his services to Theoretical Physics and especially for his discovery of the law of photoelectric effect." This well-written biography enables readers to develop an appreciation and an understanding of Einstein's work. Black and white photographs, a chronology, a bibliography, and an index are included.

Sherrow, Victoria. *Linus Pauling: Investigating the Magic Within.* Austin, Tex.: Raintree Steck-Vaughn, 1997. 112 pp. (hc. 0-8172-4400-X) $27.00.
Gr. 6–12. From the Innovative Minds series. Unusual facts about Pauling, such as the fact that he received his high school diploma at the age of sixty-one after having received two Nobel Prizes, help to make this a book that students read and remember. The book concludes with a glossary, books for further reading, a list of sources, and an index.

——. *Jonas Salk*. New York: Facts On File, 1993. 134 pp. (hc. 0-8160-
   2805-2) $25.00.
Gr. 6–12. From the Makers of Modern Science series. Sherrow's descriptions of the dire effects of polio help readers realize the importance
of Salk's work. She recounts the years of work conducted by Salk and
other researchers to find a vaccine to prevent polio. The story ends with
a chapter on Salk's legacy. A glossary, resources for further reading,
and an index conclude the book.

Billings, Charlene W. *Grace Hopper: Navy Admiral and Computer
   Pioneer*. Berkeley Heights, N.J.: Enslow, 1989. 128 pp. (hc. 0-
   89490-680-1) $20.95.
Gr. 7–12. Interviews with Grace Hopper, her family, and her friends as
well as primary source documents provided the author with a wealth of
information for this biography. Noted for her work in computer programming, Hopper is also remembered for being the first person to
"debug" a computer. Within the pages of this book is a photograph of
her logbook and the moth she removed from the computer. Black and
white photographs and reproductions, Hopper's resume, books for further reading, and an index are included.

Coil, Suzanne M. *Robert Hutchings Goddard: Pioneer of Rocketry and
   Space Flight*. New York: Facts On File, 1992. 134 pp. (hc. 0-8160-
   2591-6) $19.95.
Gr. 7–12. From the Makers of Modern Science series. The stories of H.
G. Wells and Jules Verne first opened Goddard's mind to space travel.
Reading *Scientific American* and *Popular Science News* while in school
provided him with current information pertinent to his scientific inter-
ests. His early interests and years of hard work earned him the title
"Father of Modern Rocketry." A glossary, resources for further reading, and an index conclude the book.

Gottfried, Ted. *Enrico Fermi*. New York: Facts On File, 1992. 128 pp.
   (hc. 0-8160-2623-8) $17.95.
Gr. 7–12. From the Makers of Modern Science series. Fermi's research
on nuclear energy is the focus of this biography. He received the Nobel
Prize for Physics for his work on artificial radioactivity. A glossary,
resources for further reading, and an index conclude the book.

Rummel, Jack. *Robert Oppenheimer: Dark Prince*. New York: Facts
   On File, 1992. 134 pp. (hc. 0-8160-2598-3) $19.95.

Gr. 7–12. From the Makers of Modern Science series. Oppenheimer was the head of the Manhattan Project, which led to the development of the atomic bomb. The epilogue summarizes Oppenhiemer's achievements noting that his most remarkable legacy is his belief in international arms control. A glossary, resources for further reading, and an index conclude the book.

St. George, Judith. *Dear Mr. Bell . . . Your Friend, Helen Keller*. New York: G.P. Putnam's Sons, 1992. 95 pp. (hc. 0-399-22337-1) $15.95.
Gr. 7–12. This biography of Helen Keller is also the story of her lifelong friendship with Alexander Graham Bell. He was well known for his work with the deaf and Keller's father sought Bell's advice on how best to help his daughter. Quotes, black and white photographs, an epilogue, a biography, and an index are included.

Spangenburg, Ray, and Diane K. Moser. *Niels Bohr: Gentle Genius of Denmark*. New York: Facts On File, 1995. 128 pp. (hc. 0-8160-2938-5) $25.00.
Gr. 7–12. From the Makers of Modern Science series. This biography introduces the physicist Neils Bohr and describes his relationships with other notables of his time such as Albert Einstein and Ernest Rutherford. He was awarded the Nobel Prize in Physics in 1922 for his work on a model of the atom. A glossary, resources for further reading, and an index conclude the book.

——. *Werner von Braun: Space Visionary and Rocket Engineer*. New York: Facts On File, 1995. 144 pp. (hc. 0-8160-2924-5) $25.00.
Gr. 7–12. From the Makers of Modern Science series. A reproduction of a manned space rocket von Braun drew as a teenager and his quotes interspersed through the text provide readers an intimate glimpse of this fascinating rocket scientist. A glossary, resources for further reading, and an index conclude the book.

Maurer, Richard. *The Wright Sister: Katharine Wright and Her Famous Brothers*. Brookfield, Conn.: The Millbrook Press, 2003. 128 pp. (hc. 0-7613-2564-6) $25.00.
Gr. 8–12. The contributions of the third member of the Wright team are chronicled in this fascinating biography of Katharine Wright, who helped manage her brothers business and personal affairs. She was the

only sibling to graduate from college and her life story tells much about the roles of women during the early twentieth-century. An author's note and an index are included.

Berlinski, David. *Newton's Gift: How Sir Isaac Newton Unlocked the System of the World*. New York: The Free Press, 2000. 217 pp. (hc. 0-684-84392-7) $24.00.
Gr. 9–12. Rather than a full biography of Newton, this book provides only brief information on Newton's life and instead focuses on most-noted accomplishments including *Principia Mathematica*. An appendix further explaining Newton's mathematical equations, a Newton chronology, and an index conclude the book.

Cutler, Alan. *The Seashell on the Mountaintop: A Story of Science, Sainthood, and the Humble Genius Who Discovered a New History of the Earth*. New York: Dutton, 2003. 228 pp. (hc. 0-525-94708-6) $23.95.
Gr. 9–12. The Catholic Church recognizes Nicolas Steno as "Blessed" (the first step to becoming a saint) and geologists recognize him as "the founder of geology." The focus of this biography is on the seashells he discovered on the mountainside and his explanation of how they got there found in his book *Prodromus*. This is a small volume that beautifully depicts the work of a man who changed existing perceptions about the formation of the earth. An epilogue, sources, and an index are provided.

Geison, Gerald L. *The Private Science of Louis Pasteur*. Princeton, N.J.: Princeton University Press, 1995. 378 pp. (hc. 0-691-03442-7) $75.00.
Gr. 9–12. Upon the death of Pasteur's last male descendant, Pasteur's notebooks, correspondence, and lecture notes became available to the public. Geison spent fifteen years reviewing these documents prior to writing this biography that describes flaws and deceptions in Pasteur's research. Photographs, appendixes, notes and sources, a bibliography, and an index are included.

Grosvenor, Edwin S., and Morgan Wesson. *Alexander Graham Bell: The Life and Times of the Man Who Invented the Telephone*. New York: Harry N. Abrams, 1997. 303 pp. (hc. 0-8109-4005-1) $45.00.

Gr. 9–12. Together, Grosvenor, Bell's great-grandson, and Wesson, a documentary filmmaker, have compiled a biography that can be used for both research and browsing. Approximately four hundred photographs, including some from the Bell family, a bibliography, and an index are included.

Hening, Robin Marantz. *The Monk in the Garden: How Gregor Mendel and His Pea Plants Solved the Mystery of Inheritance*. New York: Houghton Mifflin, 2000. 291 pp. (hc. 0-395-97765-70) $24.00.
Gr. 9–12. Upon his death most of Mendel's papers were burned. From the remaining papers and research on life in nineteenth-century Europe Hening tells the story of how Mendel's years of work were all but ignored by the scientific community. Then, in 1900 three different scientists in three different countries recognized the importance of Mendel's work. An epilogue, notes, selected readings, and an index are included.

Hickam, Jr. Homer H. *Rocket Boys: A Memoir*. New York: Delacorte Press, 1998. 368 pp. (hc. 0-385-33320-X) $23.95; (pbk. 0-385-33321-8) $13.95.
Gr. 9–12. A retired NASA engineer, Hickam tells about his life growing up in a coal town in West Virginia. Inspired by Russia's launch of Sputnik, Hickam and his friends begin designing and launching rockets. This venture enables Hickam to escape the West Virginia coal mines and shows that with determination amazing things are possible.

Labouisse, Eve Curie. *Madame Curie: A Biography*. New York: De Capo Press, 2001. 392 pp. (pbk. 0-306-81038-7) $17.00.
Gr. 9–12. Originally published in 1937, this biography was lovingly written by Madame Curie's younger daughter, Eve. She paints a portrait of an extraordinary woman, obsessed with her work, who eventually won two Nobel Prizes.

Maddox, Brenda. *Rosalind Franklin: The Dark Lady of DNA*. New York: HarperCollins, 2002. 380 pp. (hc. 0-06-018407-8) $29.95.
Gr. 9–12. Franklin's photographs of DNA were used by Watson and Crick when they uncovered the double helical structure of DNA, yet her contributions were largely ignored. Black and white photographs, Franklin's personal correspondence, and interviews with her friends and colleagues enhance this examination of the life and research of a

noted scientists who died at the age of thirty-six. An epilogue, notes, a bibliography, and an index conclude the book.

Masini, Giancarlo. *Marconi*. New York: Marsilio, 1995. 370 pp. (pbk. 1-56886-057-9) $14.95.
Gr. 9–12. This is a thoroughly researched account of the life and work of Marconi, whose invention of wireless transmissions continues to impact communications today. An index concludes the book.

Mead, Clifford, and Thomas Hager. *Linus Pauling: Scientist and Peacemaker*. Corvallis, Oreg.: Oregon State University Press, 2001. 272 pp. (hc. 0-87071-489-9) $35.00.
Gr. 9–12. This volume is a collection of Pauling's writings and the writings of his colleagues, friends, and students to commemorate his one-hundredth birthday. It is a very readable tribute to the work of Pauling as a scientist and peace activist. A Pauling bibliography and an index conclude the book.

Morell, Virginia. *Ancestral Passions: The Leakey Family and the Quest for Humankind's Beginnings*. New York: Simon & Schuster, 1995. 639 pp. (hc. 0-684-80192-2) $30.00; (pbk. 0-684-82470-1) $16.00.
Gr. 9–12. Family memories, letters, journals, and photographs were shared with the author as she researched the lives of the Leakey family members: Louis, Mary, and Richard. This family of paleoanthropologists worked throughout East Africa and found invaluable information about human origins. Notes, a bibliography, and an index are included.

Peat, F. David. *In Search of Nikola Tesla*. London: Ashgrove, 2003. 159 pp. (pbk. 1-85398-117-6) $14.95.
Gr. 9–12. Peat recounts Tesla's famous battle with Edison over alternating current rather than direct current. This biography describes Tesla's many inventions and their impact on society. Black and white photographs and an index are included.

Quinn, Susan. *Marie Curie: A Life*. New York: Simon & Schuster, 1995. 509 pp. (hc. 0-671-67542-7) $30.00.
Gr. 9–12. This is a well-researched biography of a scientist noted for trying to keep her public research and her private life separate. Quinn has drawn from documents released to the public in 1990 to write this compelling portrait of a complex woman. Black and white photographs, notes, a bibliography, and an index are included.

Sobel, Dava. *Longitude: The True Story of a Lone Genius Who Solved the Greatest Scientific Problem of His Time.* New York: Walker, 1995. 184 pp. (hc. 0-8027-1312-2) $19.00; (pbk. 0-14-025879-5) $11.95.
Gr. 9–12. The search for a way to determine longitude involved astronomy, clocks, navigation, politics, foul play, scientific discovery, ambition, persistence, and perfection. Where astronomers failed, clockmaker John Harrison succeeded after many long years of toil and frustration. A list of sources and an index conclude the book.

Sobel, Dava, and William J. H. Andrewes. *The Illustrated Longitude: The True Story of a Lone Genius Who Solved the Greatest Scientific Problem of His Time.* New York: Walker, 1998. 216 pp. (hc. 0-8027-1344-0) $32.95.
Gr. 9–12. This is the same story of John Harrison as noted in the above annotation; however, this one has maps, diagrams, and photographs. A bibliography and an index are included.

Strathern, Paul. *Hawking and Black Holes.* New York: Doubleday, 1998. 112 pp. (pbk. 0-385-49242-1) $15.00.
Gr. 9–12. From The Big Idea series. Strathern explains black holes in terms understandable to nonscientists and describes the life of Stephen Hawking. A timeline and suggestions for further reading are included.

——. *Turing and the Computer.* New York: Doubleday, 1999. 112 pp. (pbk. 0-385-49243-X) $15.00.
Gr. 9–12. From The Big Idea series. This is a succinct description of Alan Mathison Turing's theory of computers, his breaking of the German Military Enigma Code in World War II, and his work on artificial intelligence. There is also information about his personal life including his indecency trial. A timeline and suggestions for further reading are included.

Brands, H. W. *The First American: The Life and Times of Benjamin Franklin.* New York: Doubleday, 2000. 759 pp. (hc. 0-385-49382-2) $35.00.
Gr. 10–12. Benjamin Franklin is perhaps the most noted and written about figure in Colonial America. A man of humble origins, he achieved more than any other citizen of the land. He was a scientist, an economist, an entrepreneur, a cartographer, an inventor, and a brilliant

"thinker." Brands brings Franklin to life in this exciting biography. The book includes source notes and an index.

Israel, Paul. *Edison: A Life of Invention.* New York: John Wiley and
    Sons, 1998. 552 pp. (hc. 0-471-52942-7) $30.00.
Gr. 10–12. This well-researched biography clearly details Edison's accomplishments. He portrays Edison as the consummate experimenter who worked hard at his inventions and improved the inventions of others. An epilogue, notes, and an index conclude the book.

Lear, Linda. *Rachel Carson: Witness for Nature.* New York: Henry
    Holt, 1997. 634 pp. (hc. 0-80503427-7) $35.00.
Gr. 10–12. Carson's papers and interviews with her friends and others were some of the data sources used to compile this lengthy biography. This is a tribute to Carson's creation of the environmental movement that is helping to preserve the Earth. Black and white photographs, an afterword, a list of abbreviations, notes, a bibliography, and an index are provided.

Sobel, Dava. *Galileo's Daughter: A Historical Memoir of Science,*
    Faith, and Love. New York: Walker, 1999. 432 pp. (hc. 0-8027-
    1343-2) $27.00; (pbk. 0-1402-8055-3) $14.00.
Gr. 10–12. This biography of Galileo is drawn from one hundred twenty-four letters written to him by his eldest daughter Suor Maria Celeste, a cloistered nun. The letters reveal their loving relationship and Suor Maria Celeste's steadfast loyalty to her father when he was on trial as a heretic. Black and white reproductions, a chronology, a chart of Florentine weights, measures, and currency, a bibliography, notes, and an index are included.

Watson, James. *Genes, Girls and Gamow: After the Double Helix.* New
    York: Alfred A. Knopf, 2002. 304 pp. (hc. 0-375-41283-2) $24.00.
Gr. 10–12. Watson takes readers along and introduces them to his colleagues as he continues his research on DNA. This autobiography begins with a cast of characters and ends with a collection of George Gamow memorabilia. A Russian theoretical physicist, Gamow was also studying RNA and some of his notes and correspondence are in the memorabilia.

White, Michael. *Isaac Newton: The Last Sorcerer.* Reading, Mass.:
    Addison–Wesley, 1997. 402 pp. (hc. 0-201-48301-7) $27.00.

Gr. 10–12. Newton's insatiable curiosity led him to noteworthy scientific achievements and to the study of alchemy, biblical chronology, natural magic, and astrology. White contends that Newton's studies of science and alchemy were linked and that exploring one helped him explore the other. Black and white reproductions, references, and an index are included.

Chapter 2

Collective Biographies

These collective biographies contain brief profiles of the lives of groups of discoverers and inventors categorized by ethnicity, gender, age, or area of research. For example, there are collective biographies of African Americans, Hispanic-Americans, women, children, geologists, archeologists, and astronomers. Some of these biographies depict the struggles of minorities to have their work accepted and respected. The biographies help readers understand and appreciate the important contributions made by diverse groups of discoverers and inventors.

Amram, Fred M. B. *African American Inventors*. Mankato, Minn.: Capstone Press, 1996. 48 pp. (hc. 1-56065-361-2) $22.60.
Gr. 1–3. From the Capstone Short Biographies series. The inventors included in this easy to read book are Lonnie Johnson, Fredrick McKinley Jones, Marjorie Stewart Joyner, Elijah McCoy, and Garrett Augustus Morgan. The book begins with a brief description of the importance of inventions and patents. Photographs of the inventors and diagrams of their inventions accompany the succinct text. A glossary, resources for learning more, and an index are included.

St. John, Jetty. *African American Scientists*. Mankato, Minn.: Capstone Press, 1996. 48 pp. (hc. 1-56065-358-2) $22.60.
Gr. 3–5. From the Capstone Short Biographies series. Photographs accompany the short biographies of scientists: Franklyn G. Prendergast, Larry Shannon, Reatha Clark King, Robert Jones, and Walter Massey. A glossary, resources for learning more, useful addresses, and an index conclude the book.

——. *Hispanic Scientists*. Mankato, Minn.: Capstone Press, 1996. 48 pp.
    (hc. 1-56065-360-4) $22.60.
Gr. 3–5. From the Capstone Short Biographies series. Included in this
book are biographies of scientists: Carlos A. Ramirez, biomedical engineer; Ellen Ochoa, astronaut; Eloy Rodriguez, zoologist; Lydia Villa-
Komaroff, neurologist; and Maria Elena Zavala, botanist. Photographs
accompany the biographies. A glossary, resources for learning more,
useful addresses, and an index conclude the book.

——. *Native American Scientists*. Mankato, Minn.: Capstone Press,
    1996. 48 pp. (hc. 1-56065-359-0) $22.60.
Gr. 3–5. From the Capstone Short Biographies series. This book contains a collection of brief biographies of scientists: Fred Begay, Wilfred
F. Denetclaw, Jr., Frank C. Dukepoo, Clifton Poodry, and Jerrel Yakel.
These narratives are well-written, easy to read, and each biography is
accompanied by photographs of the scientists. A glossary, resources for
learning more, useful addresses, and an index conclude the book.

Sandler, Martin W. *Inventors*. New York: HarperCollins Children's
    Books, 1996. 93 pp. (hc. 0-06-024923-4) $21.95.
Gr. 3–12. From A Library of Congress Book series. Short bursts of text
accompanied by quotes, photographs, and reproductions describe inventors and inventions that changed the way we live. This is a brief
introduction to a wide variety of inventions and it is a great place to
begin an examination of inventors and their inventions. Information on
the Library of Congress and an index concludes the book.

Carruthers, Margaret W., and Susan Clinton. *Pioneers of Geology: Dis-
    covering Earth's Secrets*. New York: Franklin Watts, 2001. 144
    pp. (hc. 0-531-11364-7) $20.00.
Gr. 4–8. From the Lives in Science series. The work of the six geologists profiled in this book includes these areas: unconformity, uni-
formitarianism, landforms, the theory of continental drift, seafloor
spreading, and astrogeology. Black and white photographs, a geological
time scale, information on the age and the structure of the earth, a glossary, bibliography, Internet resources, and an index are included.

Casey, Susan. *Women Invent!: Two Centuries of Discoveries That
    Have Shaped Our World*. Chicago: Chicago Review Press, 1997.
    144 pp. (pbk. 1-55652-317-3) $14.95.

Gr. 4–8. These short biographies tell how the women invented things they discovered a need for just as male inventors do, however, the inventions of females are often overlooked. The book includes a resource section with information on competitions and camps for young inventors.

Cox, Clinton. *African American Healers*. New York: John Wiley and
Sons, 2000. 164 pp. (hc. 0-471-24650-6) $22.95.
Gr. 4–8. From the Black Stars Biography series. The book is divided into four sections: the early years, the Civil War years, the new century, and modern times. Each section contains brief biographies noting the outstanding contributions of several individuals; for example, William Hinton, who invented the Hinton test for syphilis. Black and white photographs, sidebars, chronology, notes, a bibliography, and an index are included.

Greenberg, Lorna, and Margot F. Horwitz. *Digging Into the Past: Pioneers of Archeology*. New York: Franklin Watts, 2001. 128 pp.
(hc. 0-531-11857-6) $20.00.
Gr. 4–8. From the Lives in Science series. Readers learn about archeologists and their work as they travel with them to uncover the secrets of King Tutankhamen's Tomb, the Indus Valley Civilization, the burial site of the sons of Ramesses II, and other sites of archeological importance. A glossary, a selected bibliography, resources for more information, and an index conclude the book.

Hatt, Christine. *Scientists and Their Discoveries*. New York: Franklin
Watts, 2001. 62 pp. (hc. 0-531-14614-6) $23.00.
Gr. 4–8. From the Documenting History series. Reproductions of primary source documents provide a personal, intimate glimpse into the lives and work of the scientists. Clear, succinct text, photographs, and diagrams help students understand the groundbreaking discoveries made by the scientists in the areas of astronomy, chemistry, physics, geology, and biology. A glossary and an index conclude the book.

Leuzzi, Linda. *Life Connections: Pioneers in Ecology*. New York:
Franklin Watts , 2000. 128 pp. (hc. 0-531-11566-6) $25.00.
Gr. 4–8. From the Lives in Science series. Eight scientists who were instrumental in developing the study of ecology are profiled in this book. There are brief biographies of Alexander von Humboldt, Jacques-Yves Cousteau, E. Lucy Braun, Aldo Leopold, Ann Haven

Morgan, Rachel Carson, Eugene Odum, and F. Sherwood Roland. The
book includes a bibliography, Internet sources, and an index.

Sullivan, Otha Richard. *African American Women Scientists and In-
    ventors*. New York: John Wiley and Sons, 2002. 150 pp. (hc. 0-
    471-38707-X) $22.95.
Gr. 4–8. From the Black Stars Biography series. These outstanding
scientists and inventors are grouped into three categories: the early
years, the new century, and modern times. These women serve as role
models for the next generation of women scientists and inventors. Their
stories are filled with courage, perseverance, and hard work. Black and
white photographs, sidebars, a chronology, notes, a bibliography, and
an index are included.

McKissack, Patricia, and Fredrick McKissack. *African American Sci-
    entists*. Brookfield, Conn.: The Millbrook Press, 1994. 96 pp. (hc.
    0-56294-372-3) $26.90.
Gr. 4–12. From A Proud Heritage series. Both well-known and lesser-
known African American scientists are profiled in this book. The book
focuses on the things of interest to each scientist and how these interests focused their research. It illustrates how scientists of different eth-
nic backgrounds contributed to discoveries, often working along similar
paths at the same time. This book concludes with a bibliography and an
index.

Mulcahy, Robert. *Diseases: Finding the Cure*. Minneapolis: The Oliver
    Press, 1996. 144 pp. (hc. 1-881508-28-5) $20.50.
Gr. 5–8. From the Innovators series. Profiled in this book are the works
of James Lind, Edward Jenner, Louis Pasteur, Frederick Banting, Alexander Fleming, and Jonas Salk. Black and white photographs, a list of
Nobel Prize winners in Physiology or Medicine, a glossary, a bibliography, and an index are included.

——. *Medical Technology: Inventing the Instruments*. Minneapolis: The
    Oliver Press, 1997. 144 pp. (hc. 1-881508-34-X) $20.50.
Gr. 5–8. From the Innovators series. Here are the stories of seven scientists whose inventions have changed medical practice. While the
chapters focus on only one scientist, they also provide information on
how each scientist built on the work of other scientists. The book concludes with an afterword, a medical timeline, the periodic table of elements, a glossary, a bibliography, and an index.

Streissguth, Thomas. *Communications: Sending the Message*. Minneapolis: The Oliver Press, 1997. 144 pp. (hc. 1-881508-41-2) $20.50.
Gr. 5–8. From the Innovators series. Scientists and inventors whose work led to the development of our present means of communication include Johann Gutenberg, Samuel Morse, Alexander Graham Bell, Thomas Edison, Guglielmo Marconi, Lee de Forest, Edwin Howard Armstrong, and Philo Farnsworth. This book contains a succinct examination of the research and the work of these men and the work of others. A communications timeline, a glossary, a bibliography, and an index conclude the book.

Wills, Susan. *Astronomy: Looking at the Stars*. Minneapolis: The Oliver Press, 2001. 144 pp. (hc. 1-881508-76-5) $20.50.
Gr. 5–8. From the Innovators series. Brief biological sketches of the developers of the science of astronomy tell about their lives and their discoveries. Included in this collection are Claudius Ptolemy, Nicolaus Copernicus, Tycho Brahe, Johnannes Kepler, Galileo Galilei, Isaac Newton, and William Herschel. An afterword, a glossary, a bibliography, and an index are included.

Anderson, Margaret J., and Karen F. Stephenson. *Scientists of the Ancient World*. Berkeley Heights, N.J.: Enslow, 1999. 104 pp. (hc. 0-7660-1111-9) $20.95.
Gr. 5–12. From the Collective Biographies series. Nine famous men and one famous woman who lived from around 580 B.C. to around 850 A.D. are profiled in this book about ten influential scholars who dominated the Ancient World. The book begins with Pythagoras and ends with Hypatia and Al-Khwarizmi. The book concludes with chapter notes and an index.

Camp, Carole Ann. *American Astronomers: Searchers and Wonderers*. Berkeley Heights, N.J.: Enslow, 1996. 104 pp. (hc. 0-89490-631-3) $20.95.
Gr. 5–12. From the Collective Biographies series. Catch the excitement of each of these astronomers as they probe the heavens. The book contains brief biographical sketches of Maria Mitchell, Percival Lowell, Williamina Fleming, Annie Jump Cannon, George Ellery Hale, Harlow Shapley, Edwin Hubble, Cecilia Payne-Gaposchkin, Vera Rubin, and Carl Sagan. Chapter notes and an index conclude the book.

French, Laura. *Internet Pioneers: The Cyber Elite*. Berkeley Heights, N.J.: Enslow, 2001. 112 pp. (hc. 0-7660-1540-8) $20.95.
Gr. 5–12. From the Collective Biographies series. French profiles the lives of ten men and women who influenced the development of the Internet: Andrew Grove, Lawrence Ellison, Ann Winblad, Esther Dyson, Steve Jobs, William H. Gates, Steve Case, Jeffrey P. Bezos, Jerry Yang, and Linus Torvalds. From software to hardware to e-commerce, these people have recognized their potential in a dynamic field. The book has chapter notes, books for further reading, Internet addresses, and an index.

Stanley, Phyllis M. *American Environmental Heroes*. Berkeley Heights, N.J.: Enslow, 1996. 127 pp. (hc. 0-89490-630-5) $20.95.
Gr. 5–12. From the Collective Biographies series. The ten American environmentalists profiled in this book are Henry David Thoreau, John Muir, Ellen Swallow Richards, George Washington Carver, Aldo Leopold, Rachel Carson, David Brower, Barry Commoner, Sylvia Earle, and Frances Moore Lappé. The book concludes with a guide to national parks, chapter notes, books for further reading, and an index.

Aaseng, Nathan. *Black Inventors*. New York: Facts On File, 1997. 128 pp. (hc. 0-8160-3407-9) $25.00.
Gr. 6–12. From the American Profiles series. Lewis Temple, Norbert Rillieux, Elijah McCoy, Lewis Latimer, Jan Matzeliger, Granville Woods, Sarah Breedlove McWilliams Walker, Garrett Morgan, Fredrick Jones, and Percy Julian and their inventions are profiled in this book. Black and white photographs of the inventors and diagrams of their inventions enhance the concise text. Each brief biography ends with a chronology and a list of books for further reading. An index concludes the book.

——. *Twentieth-Century Inventors*. New York: Facts On File, 1991. 132 pp. (hc. 0-8160-2485-5) $25.00.
Gr. 6–12. From the American Profiles series. Concise text, black and white photographs, chronologies, and books for further reading are included in the brief chapters that describe the lives and work of a variety of twentieth-century inventors who have changed our lives. Included in the book are the inventors who created the airplane, the rocket, the television, the cyclotron, the computer, the transistor, the implantable pacemaker, and the laser. An index concludes the book.

Altman, Linda Jacobs. *Women Inventors*. New York: Facts On File, 1997. 118 pp. (hc. 0-8160-3385-4) $25.00.
Gr. 6–12. From the American Profiles series. In this book readers find brief biographies on Amanda Theodosia Jones, Carrie Everson, Sara Josephine Baker, M.D., Madam C. J. Walker, Ida Rosenthal, Katharine Blodgett, Elizabeth Hazen, Rachel Brown, Bette Graham, and Ruth Handler. Black and white photographs, diagrams, and sidebars are included. Each brief biography ends with a chronology and a list of books for further reading. An index concludes the book.

Haskins, Jim. *Outward Dreams: Black Inventors and Their Inventions*. New York: Walker, 2003. 112 pp. (pbk. 0-8027-7673-6) $8.95.
Gr. 6–12. Henry Blair, Lewis Temple, James Forten, Henry Boyd, and Benjamin Bradley are African American inventors whose names may be unfamiliar to readers. These inventors and better-known ones including Benjamin Banneker, Norbert Rillieux, Elijah McCoy, and Garrett Morgan are profiled in this absorbing book. Haskins vividly describes the obstacles and racial prejudice faced by these inventors. The book concludes with a list of inventions by African Americans from 1834 to 1900, a bibliography, and an index.

Karnes, Frances A., and Suzanne M. Bean. *Girls and Young Women Inventing: Twenty True Stories about Inventors Plus How You Can Be One Yourself*. Minneapolis: Free Spirit Publishing, 1995. 168 pp. (pbk. 0-915793-89-X) $12.95.
Gr. 6–12. The first part of the book contains interesting first person accounts of young inventors and their inventions grouped in categories including convenience, work-saving, conservation, fun, health, and safety. Part two of the book contains information on how to be an inventor and part three of the book includes resources for further inspiration. An index concludes the book.

Olensky, Walter. *Hispanic-American Scientists*. New York: Facts On File, 1998. 120 pp. (hc. 0-8160-3404-3) $25.00.
Gr. 6–12. From the American Profiles series. The book includes brief chapters on Luis Alvarez, David Cardus, Manuel Cardona, Pedro Sanchez, Mario Molina, Henry Diaz, Francisco Dallmeier, Adriana Ocampo, Margarita Colmenares, and Ellen Ochoa. These scientists include physicists, a physician, a soil scientist, an environmental chemist, a meteorologist, a biologist, a geologist, and engineers. Each

chapter concludes with a short chronology of the inventor's life and a list of books for further reading. An index is included.

Stille, Darlene R. *Extraordinary Women of Medicine*. New York: Children's Press, 1997. 288 pp. (hc. 0-516-20307-X) $39.00; (pbk. 0-516-26145-2) $16.95.
Gr. 6–12. From the Extraordinary People series. Here are the stories of over fifty extraordinary women. While not all of them were inventors or discoverers, they all made lasting contributions to the field of medicine. The book includes a glossary, resources for learning more, and an index.

——. *Extraordinary Women of Science*. New York: Children's Press, 1995. 208 pp. (hc. 0-516-00585-5) $39.00.
Gr. 6–12. From the Extraordinary People series. The introduction tells readers that the scientists profiled in this book are role models for young women and they serve as evidence that women are well suited to working in the fields of math and science. There are biographies of fifty women scientists and articles that summarize the contributions of women in medicine, mathematics, and engineering. A bibliography and an index conclude the book.

Yount, Lisa. *Asian American Scientists*. New York: Facts On File, 1998. 112 pp. (hc. 0-8160-3756-6) $25.00.
Gr. 6–12. From the American Profiles series. Included in this collection are profiles of ten Asian American scientists who spent their early lives in Asia. These scientists place a high priority on education, self-discipline, precision, and achievement. Each brief biographical entry includes a photograph of the scientist, a chronology, and books for further reading.

Currie, Stephen. *Women Inventors*. San Diego: Lucent, 2001. 96 pp. (hc. 1-56006-865-5) $27.45.
Gr. 7–12. From the History Makers series. An introduction briefly describes the obstacles women inventors have overcome in order to have their work represented. The inventions and work of Temple Grandin, Madam C. J. Walker, Rose O'Neil, Grace Hopper, and Margaret Knight are included in this slim volume. Black and white photographs accompany the succinct text. Books for further reading, a bibliography, and an index are included.

Evernden, Margery. *The Experimenters: Twelve Great Chemists.* Greensboro, N.C.: Avisson Press, 2001. 200 pp. (pbk. 1-888105-49-6) $19.95.
Gr. 7–12. From the Avisson Young Adult series. Included in this book are brief biological sketches of Antoine Lavoisier, Joseph Priestley, John Dalton, Jons Jacob Berzelius, Justus von Liebig, Dmitri Mendeleyev, Louis Pasteur, Marie and Irene Curie, Richard Willstatter, Dorothy Crowfoot Hodgkin, and Linus Pauling. This book is a wonderful introduction to the work of this select group of chemists. Black and white drawings of the chemists, a bibliography, and an index are included.

Nardo, Don. *Scientists of Ancient Greece.* San Diego: Lucent, 1999. 127 pp. (hc. 1-56006-362-9) $27.45.
Gr. 7–12. From the History Makers series. Beginning with an overview of Greek Science this book is a collection of brief biographies about seven notable scientists. Brief biographies of the following Greek thinkers are included: Democritus, Plato, Aristotle, Theophrastus, Archimedes, Ptolemy, and Galen. The book includes notes, a chronology, books for further reading, a bibliography, and an index.

Sullivan, Otha Richard. *African American Inventors.* New York: John Wiley and Sons, 1998. 164 pp. (hc. 0-471-14804-0) $19.95.
Gr. 7–12. From the Black Stars series. Some twenty-five African American inventors who made significant scientific contributions dating from the 1700s to modern times are profiled. Famous persons such as George Washington Carver and Madam C. J. Walker are discussed as well as many lesser-known figures. The book includes a chronology, notes, a bibliography, and an index.

Tucker, Tom. *Brainstorm!: The Stories of Twenty American Kid Inventors.* Illus. by Richard Loehle. New York: Farrar, Straus and Giroux, 1995. 148 pp. (hc. 0-374-30944-2) $15.00.
Gr. 7–12. This interesting book for young people covers twenty inventions thought of by young Americans from ages eight to nineteen such as the Popsicle and earmuffs. Some of the children became famous, such as Thomas Edison and Benjamin Franklin. Each story is told in an easy to read style and shows the ingenuity of young people. There is a glossary and a list of sources.

Vare, Ethlie Ann, and Greg Ptacek. *Women Inventors and Their Discoveries*. Minneapolis: The Oliver Press, 1993. 160 pp. (hc. 1-881508-06-4) $19.95.

Gr. 7–12. Ten women inventors and their inventions and/or discoveries are surveyed in this book. Women profiled in the book include Elizabeth Lucas Pinckney, Martha Coston, Fannie Merritt Farmer, Madam C. J. Walker, S. Josephine Baker, Barbara McClintock, Bette Nesmith Graham, Grace Hopper, Ruth Handler, and Stephanie Kwolek. The book includes photographs, a bibliography, and an index.

Brown, David E. *Inventing Modern America: From the Mouse to the Microwave*. Cambridge, Mass.: MIT Press, 2003. 209 pp. (pbk. 0-262-52349-3) $19.95.

Gr. 9–12. Brown profiles thirty-five inventors and their inventions from the past century. George Washington Carver, Henry Ford, Stephanie Kwoleck, and Steve Wozniak are some of the inventors whose stories intrigue readers and provide role models for young inventors. An index is included.

Greenstein, George. *Portraits of Discovery: Profiles in Scientific Genius*. New York: John Wiley and Sons, 1998. 232 pp. (hc. 0-471-19138-8) $24.95.

Gr. 10–12. Presented in this book are profiles of the lives and achievements of ten distinguished scientists: Annie Jump Cannon, Cecilia Payne-Gaposchkin, Ludwig Boltzmann, George Gamow, Homi Bhabha, Luis Alvarez, Richard Feynman, Martin Perl, Margaret Geller, and John Huchra. A bibliography and an index are included.

Chapter 3

Picture Books

Picture books span genres and the illustrations either dominate the text or contribute equally to the story. The books in this chapter include picture biographies, books on construction, books on modes of transportation, and a book on the history of counting. All of the books focus on a discoverer, a discovery, an inventor, or an invention, many of which are already familiar to young readers.

Burton, Virginia Lee. *Mike Mulligan and His Steam Shovel.* Boston: Houghton Mifflin, 1939. 32 pp. (hc. 0-395-16961-5) $15.00; (pbk. 0-395-25939-8) $5.95.
Gr. P–1. Digging the cellar for Popperville's new town hall in one day seems like an impossible feat, but Mike and his steam shovel, Mary Anne, prove it can be done. Generations of youngsters have grown up with this beloved tale of a machine and its determined owner.

Crews, Donald. *Freight Train.* New York: Greenwillow Books, 1978. 32 pp. (hc. 0-688-80165-X) $15.99.
Gr. P–1. Illustrations of bright, colorful train cars introduce a freight train to the youngest readers. Once all the cars are introduced the freight train speed across the pages through tunnels and over trestles.

——. *School Bus.* New York: Greenwillow Books, 1984. 32 pp. (hc. 0-688-02807-1) $15.99.
Gr. P–1. Short, simple sentences in large, dark print at the bottom of the pages describe the school bus's journey from morning until early evening. Colorful pictures filled with bright yellow school buses hold young readers' attention.

Collicutt, Paul. *This Boat*. New York: Farrar Straus and Giroux, 2001.
32 pp. (hc. 0-374-37495-3) $15.00.
Gr. P–2. Simple text and dynamic, colorful illustrations present a basic
introduction to a wide variety of boats. This is a captivating book that
young readers return to repeatedly. The endpapers contain drawings of
a historical collection of boats from different countries.

——. *This Car*. New York: Farrar Straus and Giroux, 2002. 32 pp. (hc.
0-374-39965-4) $15.00.
Gr. P–2. Bight, colorful illustrations highlight a wide variety of cars
from lunar rovers to jeeps to roadsters. Short simple sentences under
the illustrations provide a lesson in opposites such as a solar-powered
car and a gas-powered car. The book is a delight to read and one that is
just right for reading more than once.

——. *This Plane*. New York: Farrar Straus and Giroux, 2002. 32 pp.
(pbk. 0-374-47517-2) $5.95.
Gr. P–2. Aviation history unfolds across the pages of this introductory
text in dramatic, colorful illustrations. One-sentence descriptors are
found on each page. The endpapers contain drawings of planes with the
country of origin and other information noted.

——. *This Train*. New York: Farrar Straus and Giroux, 2001. 32 pp. (hc.
0-374-37493-7) $15.00; (pbk. 0-374-47518-0) $5.95.
Gr. P–2. The illustrations feature brightly colored trains zipping across
the countryside. One-sentence descriptors of opposites appear at the
bottom of the illustrations. The endpapers contain drawings of various
types of trains from different countries.

Hunter, Ryan Ann. *Dig a Tunnel*. Illus. by Edward Miller. New York:
Holiday House, 1999. 32 pp. (hc. 0-8234-1391-8) $15.95.
Gr. P–2. Tunnels enable animals, people, and goods to move under the
ground and beneath the water. Colorful illustrations and brief text explain how tunnels are constructed and how they are used. Included in
the book are descriptions of the Chesapeake Bay Bridge-Tunnel, the
Chunnel, and the Mont Blanc Car Tunnel.

Joyce, William. *Sleepy Time Olie*. New York: HarperCollins, 2001. 32
pp. (hc. 0-06-029613-5) $15.95.
Gr. P–2. Futuristic robot Olie is ready to go to sleep but Pappy is not
home. When Pappy comes home squeaking, frowning, and grumpy,

Olie invents a "super silly ray" that restores Pappy's good humor and the household goes peacefully to sleep. Vibrant, colorful, computer-enhanced illustrations hold readers' attention as does the rhyming text.

Konigsburg, E. L. *Samuel Todd's Book of Great Inventions*. New York: Atheneum, 1991. 32 pp. (hc. 0-689-31690-1) $13.95.
Gr. P–2. From Samuel Todd's perspective the really great inventions are those that make his world a better place, such as a thermos bottle that keeps Kool-Aid cool or a backpack to keep his hands free. The first invention mentioned is a mirror and the back cover of the book is a mirror image of the front cover. This is a great introduction to inventions for young readers.

Marshall, Edward. *Space Case*. Illus. by James Marshall. New York: Dial Books for Young Readers, 1980. 32 pp. (hc. 0-8037-8007-9) $14.99.
Gr. P–2. When a thing arrives from outer space on Halloween, it joins right in with a group of children as they trick-or-treat in their neighborhood. At the end of the night, it follows one of the children home and on to school the next day. The adults in the story seem oblivious to the fact that this in no ordinary creature. The faces on the characters in the amusing illustrations capture the deadpan humor of the text.

Chall, Marsha Wilson. *Prairie Train*. Illus. by John Thompson. New York: HarperCollins, 2003. 32 pp. (hc. 0-688-13433-5) $15.99.
Gr. P–3. A young girl experiences her first train ride when she goes from her home in the country to visit her grandmother in the city. It is in the first half of the twentieth-century that she rides on the Great Northern. Her loneliness on the trip is broken up by her visit to the dining car, a young boy playing the harmonica for her to sing along with, and by a lovely older passenger showing her how to knit. This is a good first look at trains for readers who rarely see passenger trains.

Hopkinson, Deborah. *Maria's Comet*. Illus. by Deborah Lanino. New York: Atheneum Books for Young Readers, 1999. 32 pp. (hc. 0-689-81501-8) $16.00.
Gr. P–3. Entranced by the night sky, Maria Mitchell joins her father on the roof as he sweeps the sky with his telescope. Dark illustrations filled with twinkling stars enfold readers into the wonders of the heavens as they learn about a young girl's dream to study the stars. Maria

Mitchell became America's first woman astronomer. An author's note and a brief glossary conclude the book.

Seymour, Tres. *Our Neighbor is a Strange, Strange Man*. Illus. by Walter Lyon Krudop. New York: Orchard, 1999. 28 pp. (hc. 0-531-33107-5) $16.99.
Gr. P–3. Most people believe that Orville and Wilbur Wright invented the first airplane. When in fact, Melville Murrell received a patent for "The Great American Flying Machine" in 1877, when Orville was six-years-old and Wilbur was ten-years-old. A repeating refrain and large colorful pictures hold the attention of young readers as they learn about Melville and his flying machine.

Wallner, Alexandra. *The First Air Voyage in the United States: The Story of Jean-Pierre Blanchard*. New York: Holiday House, 1996. 32 pp. (hc. 0-8234-1224-5) $15.95.
Gr. P–3. Jean-Pierre Blanchard flew from Philadelphia to Woodbury, New Jersey in a hot air balloon in 1793. Since he spoke only French, he carried a letter of introduction from President George Washington on his journey. When he landed in New Jersey, the people read the letter and helped him get back to Philadelphia. Located at the end of the book are an author's note and a pronunciation guide for the French words found in the book.

Compestine, Ying Chang. *The Story of Paper*. Illus. by Yong Sheng Xuan. New York: Holiday House, 2003. 32 pp. (hc. 0-8234-1705-0) $16.95.
Gr. K–2. As in the stories about chopsticks, noodles, and kites, the Kang family is involved in a delightful story about how paper was invented. The three boys are always in trouble at school for their insatiable curiosity. The teacher is always sending notes home to their parents–written on their hands! The boys decide to find a way to invent something else for the teacher to write on and their invention is good enough to be presented to the emperor. At the end of the book are an author's note and a recipe for homemade garden paper for the students to try.

Edwards, Pamela Duncan. *The Wright Brothers*. Illus. by Henry Cole. New York: Hyperion Books for Children, 2003. 40 pp. (hc. 0-7868-1951-0) $15.99.

Gr. K–2. This humorous, cumulative tale explains the Wright brothers' experiments in flight to younger students. Small mice appear in the illustrations and their conversations supply additional information and humorous asides. The endpapers contain a ribbon timeline of flight.

Brown, Don. *Rare Treasure: Mary Anning and Her Remarkable Discoveries*. Boston: Houghton Mifflin, 1999. 32 pp. (hc. 0-395-92286-0) $15.00.
Gr. K–3. As a child, Mary Anning hunted fossils with her father on the beach near their house. They sold the fossils to their neighbors and tourists. When her father died, Mary helped support her mother and brother by selling her fossil finds. Over the years, Mary and her fossils became known around the world. Watercolor illustrations accompany the text.

Denslow, Sharon Phillips. *Radio Boy*. Illus. by Alec Gillman. New York: Simon & Schuster Books for Young Readers, 1995. 32 pp. (hc. 0-689-80295-1) $15.00.
Gr. K–3. This picture book biography describes the early life of Nathan B. Stubblefield and life in small town America during the 1800s. Stubblefield tinkered with "electricals" and fixed a broken telephone as a youngster. While the story itself does not include much about his invention of the radio an author's note at the end of the book tells about this invention and some of his patents.

Howard, Ginger. *William's House*. Illus. by Larry Day. Brookfield, Conn.: The Millbrook Press, 2001. 32 pp. (hc. 0-7613-1674-4) $22.90.
Gr. K–3. William and his family moved from England and settled in New England in 1637. William built a house just like the one his father had in England. Unfortunately, the design of the house was not suited for the New England climate. As each season came and went William made changes in the house to suit the New England climate, such as digging a root cellar and replacing the flat roof with a pitched roof.

Joyce, William. *A Day with Wilbur Robinson*. New York: Harper and Row, 1990. 32 pp. (hc. 0-06-022967-5) $14.95.
Gr. K–3. Spending the day with Wilbur Robinson involves a search for Grandfather's missing teeth, introductions to a variety of wacky relatives, and an odd assortment of innovative devices. The family has a robot named Carl, Uncle Judlow has a contraption for augmenting his

brain, Uncle Laszlo has a new antigravity device, and Uncle Art arrives from abroad in a space ship. The humorous illustrations contain details indicating that strange things have happened.

Martin, Jacqueline Briggs. *Snowflake Bentley*. Illus. by Mary Azarian. Boston: Houghton Mifflin, 1998. 32 pp. (hc. 0-395-86162-4) $16.00; (pbk. 0-486-41526-0) $5.59.
Gr. K–3. This picture book biography captures Bentley's passion for the beauty in snowflakes. He developed microphotography to take pictures of snowflakes and then studied them. He discovered that no two snowflakes are alike. Hand-tinted woodcut illustrations and Bentley's snowflake photographs make this book a delight to examine. This book won the Caldecott.

Atkins, Jeannine. *Mary Anning and the Sea Dragon*. Illus. by Michael Dooling. New York: Farrar Straus and Giroux, 1999. 32 pp. (hc. 0-374-34840-5) $16.00.
Gr. K–4. Subdued oil paintings capture Mary Anning at work searching for fossils on the English seashore near her home. Mary learned to dig fossils from her father. When he died, she continued digging to help support her family. This became her life work and some of her discoveries are preserved in museums. An afterword concludes the book.

Carlson, Laurie. *Boss of the Plains: The Hat That Won the West*. Illus. by Holly Meade. New York: DK Publishing, 1998. 32 pp. (hc. 0-7894-2479-7) $16.95.
Gr. K–4. John Batterson Stetson left the family hat shop in Orange, New Jersey and headed West. He created a tough felt hat to keep the sun out of his eyes and the rain from dripping down his back as he dug for gold in Colorado. When the search for gold did not prove lucrative, he moved to Philadelphia and began creating his special hat called the "Boss of the Plains." The book ends with a mini-biography of Stetson and a bibliography.

Oppenheim, Shulamith Levey. *Rescuing Einstein's Compass*. Illus. by George Juhasz. Northampton, Mass.: Interlink Publishing Group, 2003. 32 pp. (hc. 1-56656-507-3) $15.95.
Gr. K–4. When Einstein was a child and sick in bed, his father gave him a compass to entertain him. The compass fascinated Einstein who carried it with him even as an adult. In this story eight-year-old Theo

rescues the compass when it falls into the water while he and Einstein are sailing.

Towle, Wendy. *The Real McCoy: The Life of an African American Inventor.* New York: Scholastic, 1995. 32 pp. (hc. 0-87207-296-7) $13.85; (pbk. 0-590-48102-9) $5.99.
Gr. K–4. Acrylic illustrations portray the hard work, courage, and dignity of this "master mechanic and engineer" trained in Scotland who could not find work in America as an engineer. Among Elijah McCoy's many patents was one for an automatic oil cup to lubricate heavy equipment. His oil cup was superior to those made by others, hence, people asked for "the real McCoy."

Wells, Rosemary, and Tom Wells. *The House in the Mail.* Illus. by Dan Andreasen. New York: Viking, 2002. 32 pp. (hc. 0-670-03545-9) $16.99.
Gr. K–4. Emily compiles a scrapbook to record the progress of the construction of the mail-order house her parents ordered from the Sears Roebuck catalog in 1927. The scrapbook shows how the land was cleared, how the foundation was poured, and how the house was put together from the prefabricated pieces that arrived in boxes. Emily and her brother are excited about the modern conveniences including running water and an electric refrigerator.

Van Allsburg, Chris. *Zathura: A Space Adventure.* Boston: Houghton Mifflin, 2002. 32 pp. (hc. 0-618-25396-3) $18.00.
Gr. K–5. Taking up where the book *Jumanji* left off, the two Budwig brothers bring home the game they found under the tree. Wedged in the bottom of the game box is a space adventure game that takes the siblings on a wild ride through outer space. When Walter is swallowed by a black hole the boys travel back in time and once again find the game box under the tree.

Schulz, Walter A. *Will and Orv.* Illus. by Janet Schulz. Minneapolis: Carolrhoda Books, 1991. 48 pp. (hc. 0-87614-669-8) $22.60.
Gr. 1–3. From the Carolrhoda On My Own Book series. An author's note in this easy to read book explains that this is a fictionalized account of the Wright brothers' first flight as told by one of the witnesses, a young boy named Johnny Moore. The author captures the excitement and sense of adventure Johnny must have felt on this historic day.

Adler, David A. *A Picture Book of George Washington Carver*. Illus. by Dan Brown. New York: Holiday House, 1999. 32 pp. (hc. 0-8234-1429-9) $15.95.

Gr. 1–4. This brief biography of a noted scientist tells students of his important work, provides information on the racial prejudice he endured, and shows his devotion to improving the agricultural practices of poor southern farmers. The book includes colorful illustrations, a chronology, a selected bibliography, and an author's note.

Brighton, Catherine. *The Fossil Girl: Mary Anning's Dinosaur Discovery*. Brookfield, Conn.: The Millbrook Press, 1999. 32 pp. (hc. 0-7613-1468-7) $22.90.

Gr. 1–4. The comic book format appeals to young readers who will be anxious to read and learn about the amazing Mary Anning. She spent her life unearthing fossils on the seashore near her home in England. In addition, her discoveries made her internationally famous. An author's note contains additional information on Anning's life.

Brill, Marlene Targ. *Margaret Knight: Girl Inventor*. Illus. by Joanne Friar. Brookfield, Conn.: The Millbrook Press, 2001. 32 pp. (hc. 0-7613-1756-2) $22.90.

Gr. 1–4. In the mid-1800s, when the thread on a factory-weaving loom broke, the shuttle would slip often injuring the workers. At ten years of age Margaret "Mattie" Knight invented a way to have the shuttle stop when the thread broke thus preventing injury to the workers. An afterword explains to readers that throughout her lifetime Mattie continued to invent. Her most famous invention was the square bottom grocery bag. A glossary concludes the book.

Dooling, Michael. *The Great Horse-Less Carriage Race*. New York: Holiday House, 2002. 32 pp. (hc. 0-8234-1640-2) $16.95.

Gr. 1–4. In November 1895, the *Chicago Times–Herald* sponsored a race across the city of Chicago to prove that the horseless carriage was better than the horse and carriage. The participants knew that the winner of the race would be able to convince the public to buy cars produced by his company. This is an exciting tale of a race beset by snow, ice, and mechanical difficulties. Dooling's illustrations add to the suspense and excitement. An author's note concludes the book.

Kroll, Steven. *Robert Fulton: From Submarine to Steamboat*. Illus. by
 Bill Farnsworth. New York: Holiday House, 1999. 32 pp. (hc. 0-
 8234-1433-7) $16.95.
Gr. 1–4. At fourteen Fulton designed an air gun used by soldiers during
the American Revolution. Later in life, he improved on designs for the
submarine and the steamboat. While he did not invent the steamboat, he
developed one that became a profitable commercial venture. The book
includes a list of important dates and an author's note.

Schanzer, Rosalyn. *How Ben Franklin Stole the Lightning*. New York:
 HarperCollins, 2003. 32 pp. (hc. 0-688-16993-7) $16.99.
Gr. 1–4. Ben Franklin was an incredible individual, being an inventor, a
printer, an author, an athlete, a scientist, and a patriot among other
things. This delightfully illustrated book is for younger children and
will please even slightly older children as well. The book concludes
with an author's note and a list of some of Ben's scientific work.

Yolen, Jane. *My Brothers' Flying Machine*. Illus. by Jim Burke. Boston: Little, Brown, 2003. 32 pp. (hc. 0-316-97159-6) $16.95.
Gr. 1–4. Katharine Wright shares with readers the exploits of her famous brothers, who often mentioned her invaluable assistance over the
years. Colorful oil paintings accompany the lyrical and informative
text.

Geisel, Theodore Seuss. *The Butter Battle Book*. New York: Random
 House, 1984. 56 pp. (hc. 0-394-86580-4) $14.95.
Gr. 1–6. The Zooks eat their bread with the butter side down and the
Yooks eat their bread with the butter side up. This is the basis for a
feud that leads to an arms race. The book ends with a Zook and a Yook
each holding a weapon capable of destroying them all. This allegorical
tale is one to read aloud, to discuss, and to help students make connections with the world beyond their classrooms.

——. *The Lorax*. New York: Random House, 1976. 70 pp. (hc. 0-394-
 82337-0) $14.95.
Gr. 1–6. Word play, rhymes, colorful illustrations, and inimitable Dr.
Seuss characters tell of the destruction of a habitat as factories grow
and the inhabitants ignore the consequences of their actions.

St. George, Judith. *So You Want to Be an Inventor?* Illus. by David
    Small. New York: Philomel Books, 2002. 52 pp. (hc. 0-399-23593-
    0) $16.99.
Gr. 1–6. Many men and women have been known as inventors, some
famous others not so famous. We all know about the radio and the
printing press, but few of us are aware of the haircutting helmet or eyeglasses for chickens! The creators of *So You Want to Be President* have
combined again to please youngsters and older children with information about inventors and inventions. Lively caricatures fill the pages
and enhance the humorous tone of the narrative. At the end of the book,
there are biographical notes and a bibliography.

Stanley, Diane. *Leonardo da Vinci.* New York: William Morrow, 1996.
    48 pp. (hc. 0-688-10437-1) $16.95; (pbk. 0-688-16155-3) $6.99.
Gr. 1–6. Leonardo da Vinci was a painter, an engineer, a scientist, and
an inventor. This picture book biography helps readers discover the
wonder of this visionary genius. Old Master–style illustrations and
drawings from da Vinci's notebook add just the right tone to this biography of a fascinating man.

Anholt, Laurence. *Stone Girl, Bone Girl: The Story of Mary Anning.*
    Illus. by Sheila Moxley. New York: Orchard Books, 1998. 32 pp.
    (hc. 0-531-30148-6) $15.95.
Gr. 2–4. Mary Anning and her small dog were familiar sights on the
coast of Lyme Regis in Dorset, England in the 1800s as they searched
for fossils to sell in her curiosity shop. As a young girl, she discovered
an ichthyosaur fossil, which was only one of her famous finds. Folk art
illustrations and flowing narrative tell Anning's story.

Crowther, Robert. *Robert Crowther's Amazing Pop-Up House of In-
    ventions.* Cambridge, Mass.: Candlewick, 2000. 12 pp. (hc. 0-
    7636-0810-6) $15.00.
Gr. 2–4. While pop-up books may not be the sturdiest books to consider purchasing for a classroom or a library, this one is an excellent
choice to begin a study of inventions. The kitchen, the living room, the
bathroom, the bedroom, and the garage are all teaming with inventions.
Written next to each object in the rooms is a brief caption about the
invention. A timeline of household inventions concludes the book.

Lasky, Kathryn. *The Librarian Who Measured the Earth*. Illus. by Kevin Hawkes. Boston: Little, Brown, 1994. 48 pp. (hc. 0-316-51526-4) $16.95.
Gr. 2–5. Over two thousand years ago, Eratosthenes determined the circumference of the earth and recent calculations using modern technology determined that his measurement was off by only about two hundred miles. As readers discover Eratosthenes, they also meet other Greek discoverers and inventors. For example, Herophilius, who discovered the differences between arteries and veins, and Ctesibius, who invented the first water-driven clock. An author's note, an afterword, and a bibliography are included.

O'Brien, Patrick. *The Hindenburg*. New York: Henry Holt, 2000. 32 pp. (hc. 0-8050-6415-X) $17.00.
Gr. 2–5. O'Brien chronicles the development and construction of zeppelins including the most famous, the Hindenburg. A two-page spread shows a cutaway view of the Hindenburg drawn as a blueprint. The tragic last flight of the Hindenburg from takeoff to landing tells of the demise of the zeppelins. At the end of the book is a collection of fascinating facts.

Rumford, James. *Seeker of Knowledge: The Man Who Deciphered Egyptian Hieroglyphs*. Boston: Houghton Mifflin, 2000. 32 pp. (hc. 0-395-97934-X) $15.00.
Gr. 2–5. At eleven years of age, Jean-François Champollion decided that one day he would decipher the Egyptian hieroglyphs he saw on the Egyptian treasures of a scientist he met. He studied ancient languages and one day deciphered ancient writings. A pronunciation guide for the Egyptian words in the book and a brief bibliography are included.

Schmandt-Besserat, Denise. *The History of Counting*. Illus. by Michael Hayes. New York: Morrow Junior Books, 1999. 45 pp. (hc. 0-688-14118-8) $17.00.
Gr. 3–5. This is an overview of the history of counting told in a picture book format. Acrylic paintings of a variety of cultures enhance this informative, easy-to-understand text. Counting is considered by some to be the greatest invention.

Sullivan, Anne Marie. *Albert Einstein: Scientist–Theory of Relativity*. Illus. by Giuliano Ferri. Philadelphia: Mason Crest, 2003. 30 pp. (hc. 1-59084-140-9) $19.95.

Gr. 3–5. From the Great Names series. As Albert Einstein grew, his life took him to many countries. However, his mind took him even further–further than any explorer had ever traveled. He seemed to be a quiet child who grew to be a quiet man. He had messy hair and rumpled clothes. But inside his mind he soared like a bird, seeing a world others could not even imagine. He loved math and science and that helped launch us into the Atomic Age. He devoted his life to research and to world peace. The illustrations are quite wonderful and enhance the text.

Visconti, Guido. *The Genius of Leonardo*. Illus. by Bimba Landmann. New York: Barefoot Books, 2000. 36 pp. (hc. 1-84148-301-X) $16.99.
Gr. 3–5. Life with Leonardo is described by his apprentice, Giacomo. Information about Leonardo's eccentric ways, his glorious paintings, his diagrams, and his inventions are accompanied by quotes from Leonardo's writings and reproductions of his works. The book begins and ends with thumbnails of Leonardo's sketches and diagrams that display his creative genius.

Lasky, Kathryn. *The Man Who Made Time Travel*. Illus. by Kevin Hawkes. New York: Farrar, Straus and Giroux, 2003. 48 pp. (hc. 0-374-34788-3) $17.00.
Gr. 3–6. This picture book biography of John Harrison, who developed a workable way to measure longitude, also briefly describes others' failed attempts. Clear, concise prose tells the story of a persistent clockmaker who worked for years to improve his sea clock. Colorful paintings contain additional details and the endpapers depict Harrison's revolutionary H–4 sea clock. An author's note and bibliographies conclude the book.

Parker, Nancy Winslow. *Lock, Croc, and Skeeters: The Story of the Panama Canal*. New York: William Morrow, 1996. 32 pp. (hc. 0-688-12241-8) $16.00.
Gr. 3–6. A poem about the Isthmus of Panama by American poet James Stanley Gilbert with colorful cartoon-like illustrations starts the book off with a humorous look at the dangers encountered in Panama. The book then takes a serious turn as information on the construction of the Panama Canal, biographies of men involved in the construction of the canal, information on the eradication of mosquitoes, a description of a trip through the canal, and a map of the canal come together to tell the history of this monumental construction feat.

Sis, Peter. *The Tree of Life*. New York: Farrar, Straus and Giroux, 2003. 36 pp. (hc. 0-88899-564-4) $18.00.

Gr. 3–6. From schoolboy to medical student to naturalist, Sis recounts Darwin's life in this beautifully illustrated book. Diary entries, charts, maps, and a gatefold spread describe Darwin's five-year voyage on the Beagle. This voyage enabled him to explain the process of evolution worked.

Chapter 4

Fiction

Books in this chapter include selected titles of poetry, fantasy, science fiction, realistic fiction, and fictionalized biographies. Through these book readers can identify with discoverers and inventors as well as with children their own age. Time travel, bullies, and a robotic brother are some of the topics explored in these books. Included in this chapter are select novels from noted science fiction writers Ray Bradbury, Jules Verne, and H. G. Wells. The inventors' biographies annotated in chapter 1 indicate that some of the inventors read the works of Verne and Wells when they were growing up.

Cassedy, Sylvia. *Zoomrimes: Poems about Things That Go.* Illus. by Michele Chessare. New York: HarperCollins, 1993. 51 pp. (hc. 0-06-022632-3) $13.89.
Gr. K–3. This is a delightful collection of twenty-six alphabetically organized poems about things that go. Black and white illustrations add to the fun and the humor. While not all of the things that go are inventions, most of them are, such as the escalator, the subway, and the vacuum cleaner.

Perry, Andrea. *Here's What You Do When You Can't Find Your Shoe: Ingenious Inventions for Pesky Problems.* Illus. by Alan Snow. New York: Simon & Schuster, 2003. 40 pp. (hc. 0-689-83067-X) $16.95.
Gr. K–5. Clever, rhythmic poems accompanied by pen and ink illustrations introduce fantasy inventions that solve common childhood problems such as lost shoes, dropped crumbs, and yucky vegetables. The humorous poems and eye-catching artwork are a great way to spark

children's imaginations and motivate them to invent something to take care of their own pesky problems.

Goldsmith, Howard. *Thomas Edison to the Rescue.* Illus. by Anna Di-Vito. New York: Simon & Schuster, 2003. 32 pp. (hc. 0-689-85332-7) $11.89; (pbk. 0-689-85331-9) $3.99.
Gr. 1–2. From the Childhood of Famous Americans series. In simple, easy to read text, students learn how Thomas Edison saved a young boy from a runaway boxcar. The boy was the son of the railroad station's telegraph operator. This is a fictionalized account of a true incident in Edison's life and it would have a profound impact on him.

Ogren, Cathy Stefanec. *The Adventures of Archie Featherspoon.* Illus. by Jack E. Davis. New York: Aladdin, 2002. 43 pp. (hc. 0-689-84284-8) $11.89; (pbk. 0-689-84359-3) $3.99.
Gr. 1–3. From the Ready for Chapters series. This action packed tall tale is set in the Wild West and the unlikely hero of the story is a young boy, Archie. Instead of doing his farm chores, Archie spends his time inventing. His inventions enable him to send outlaws packing, to create the perfect birthday present for his mother, and eventually to get the corn planted.

Duffey, Betsy. *The Gadget War.* Illus. by Janet Wilson. New York: Viking, 1991. 75 pp. (hc. 0-670-84152-8) $13.99; (pbk. 0-14-130708-0) $4.99.
Gr. 2–4. Kelly Sparks is a third grade gadget wizard with forty-three inventions to her credit. When a new student named Albert Einstein Jones arrives, he and Kelly declare war to determine which one of them is the real gadget wizard. Throughout the text, there is information about real inventors and their inventions. The silly antics in the book appeal to young readers who are ready for short chapter books.

Pinkwater, Daniel. *Ned Feldman, Space Pirate.* New York: Macmillan, 1994. 48 pp. (hc. 0-02-744633-X) $14.95.
Gr. 2–4. This delightful and entertaining author has taken on the fantasy-lives of children to tell this humorous story. One day when Ned's parents are out, he meets Captain Lumpy Lugo, a space pirate who comes from the galaxy Foon-ping-baba and appears underneath the kitchen sink. They spend the afternoon traveling through outer space. Pinkwater's original illustrations are bright and cheery and will have special appeal to young readers.

Wishinsky, Frieda. *Manya's Dream*. Illus. by Jacques Lamontagne. Toronto, Ont.: Maple Tree Press, 2002. 32 pp. (hc. 1-894379-53-5) $19.95; (pbk. 1-894379-54-3) $6.95.
Gr. 3–4. Readers listen in as a mother tells her daughter the story of Marie Curie's life. An introduction explains the importance of Curie's discoveries. A chronology and a summary of Curie's achievements are included.

——. *What's the Matter with Albert?* Illus. by Jacques Lamontagne. Toronto, Ont.: Maple Tree Press, 2002. 32 pp. (hc. 1-894379-31-4) $19.95; (pbk. 1-894379-32-2) $6.95.
Gr. 3–4. This is a fictionalized interview-biography of Albert Einstein. A young boy, Billy, interviews Einstein for the school newspaper. Readers learn that Einstein had difficulties in school and they learn something about his discoveries, which may lead them to do additional research on Einstein. The book includes an author's note and a chronology.

Scieszka, Jon. *Hey Kid, Want to Buy a Bridge?* Illus. by Adam McCauley. New York: Viking, 2000. 74 pp. (hc. 0-670-89916-X) $14.99.
Gr. 3–5. From the Time Warp Trio series. Traveling back to 1877, the boys land on a tower of the unfinished Brooklyn Bridge and are joined by their great-granddaughters. In this action-packed adventure, they meet inventor Thomas Edison and Brooklyn Bridge engineer Washington Roebling. At the end of the book, Scieszka invites readers to invent something.

Weiss, Ellen, and Mel Friedman. *The Poof Point*. New York: Alfred A. Knopf, 1992. 166 pp. (hc. 0-679-83257-2) $14.00; (pbk. 0-679-82272-0) $3.99.
Gr. 3–5. Marigold and Norton Bickers are inventors. When their inventions do not quite work as they plan, it is up to their children, twelve-year-old Marie and nine-year-old Eddie, to fix things. This time the Bickers create a time machine that does not physically transport them back in time. They just become mentally younger and younger and are in danger of eventually ceasing to exist. Marie and Eddie come to their rescue and restore their parents to their correct ages.

Woodruff, Elvira. *The Disappearing Bike Shop*. New York: Holiday House, 1992. 169 pp. (hc. 0-8234-0933-3) $13.95.

Gr. 3–5. A strange-looking bike shop appears, disappears, and reappears at the end of Dewberry Street. Three boys venture into the shop and meet the proprietor, Mr. Quigley. When two of the boys are assigned a class report on Leonardo da Vinci, they realize that the time traveling proprietor is da Vinci. Mystery, suspense, history, and friendship are all woven into this engaging story.

Becker, Bonny. *My Brother, the Robot*. New York: Dutton Children's Books, 2001. 136 pp. (hc. 0-525-46792-0) $15.99.
Gr. 3–6. When robotic Simon, the perfect son, arrives, Chip feels he has been replaced. To save money Chip's dad bought a not-quite-perfect robot and not being quite perfect is what in the end convinces the family to keep Simon.

Getz, David. *Almost Famous*. New York: Henry Holt, 1992. 182 pp. (hc. 0-8050-1940-5) $13.95.
Gr. 4–6. Ten-year-old Maxine is convinced that one day she will become a famous inventor and appear on Phil Donahue's show. She receives an application in the mail for an invention contest, but the problem is that she needs a partner and a teacher sponsor. Undaunted, she sets out to find both and succeeds.

Haseley, Dennis. *The Amazing Thinking Machine*. New York: Dial Books, 2002. 117 pp. (hc. 0-8037-2609-0) $16.99.
Gr. 4–6. Roy and Patrick's father left home looking for work, as did many other fathers during the Depression. Roy decides that to take their minds off their troubles and to help put food on the table they are going to build an amazing thinking machine. The neighborhood children soon line up to put a penny or some food into the machine in exchange for clever answers to their questions.

Lawson, Robert. *Ben and Me: A New and Astonishing Life of Benjamin Franklin as Written by His Good Mouse Amos*. Boston: Little, Brown, 1988. 114 pp. (pbk. 0-316-51730-5) $5.99.
Gr. 4–6. Few people realize that the source of many of Benjamin Franklin's ideas for inventions came from a mouse named Amos! This extraordinary mouse provides a humorous, amusing, and witty account of these inventions and results. This book is a Caldecott Medal Winner.

Rendal, Justine. *A Very Personal Computer*. New York: HarperCollins, 1995. 216 pp. (hc. 0-06-025404-1) $16.95.

Gr. 4–6. Twelve-year-old Pollard has more than his share of problems and is considered an underachiever. When he goes to the computer lab for remedial work, he meets Compensatory Program One, Conner. This computer program provides help with his homework, virtual reality simulations, and guidance.

Weitzman, David. *Old Ironsides*. Boston: Houghton Mifflin, 1997. 32 pp. (hc. 0-395-74678-7) $15.95.
Gr. 4–6. This is a fictionalized account of the construction of the warship *Constitution*, as seen through the eyes of young John Aylwin. Pen and ink drawings show the details of the ship's construction and the innovations of designer Joshua Humphreys. An epilogue explains that in battle a shot fired at the ship's hull bounced off, hence earning the frigate the name "Old Ironsides."

——. *Thrashin' Time: Harvest Days in the Dakotas*. Boston: David R. Godine, 1991. 77 pp. (hc. 0-87923-910-7) $14.95.
Gr. 4–6. Set in the early1900s, this story describes harvest time on a family farm in North Dakota and the excitement generated by the steam-powered threshing rig. Pen and ink drawings capture this forgotten time in America's past and in combination with the text help students understand the importance of steam-powered machinery. The endpapers contain a labeled drawing of the thresher.

L'Engle, Madeleine. *A Wrinkle in Time*. New York: Yearling Books, 1973. 240 pp. (pbk. 0-4404-9805-8) $6.50.
Gr. 4–7. According to most people in town, Meg Murry is both volatile and dull-witted and her brother Charles Wallace is dumb. These same people say that their physicist father has run off, leaving their mother, a brilliant scientist, and their children alone. But a new neighbor, Mrs. Whatsit, does not agree and she helps Meg, Charles Wallace, and a new friend, Calvin O'Keefe, as they embark on a perilous quest through space to find their father by tesseracting. This is a Newbery Medal book.

Gates, Phil. *The History News: Medicine*. Cambridge, Mass.: Candlewick, 1997. 32 pp. (hc. 0-7636-0316-3) $22.60.
Gr. 4–8. Fictionalized accounts of factual medical discoveries in a newspaper format grab readers' attention. Fake advertisements, corny jokes, colorful illustrations, and sensationalized headlines complete the tabloid. An index is provided.

Gutman, Dan. *Virtually Perfect.* New York: Hyperion Books for Children, 1998. 123 pp. (hc. 0-7868-0394-0) $12.95; (pbk. 0-7868-1745-3) $5.99.

Gr. 4–8. With a retired grandfather who used to create special effects for movies and a father who creates virtual actors on the computer, Yip has plenty of opportunities to explore his own creativity. After accessing his father's computer files, Yip creates a virtual reality actor, Victor, who steps out of the computer and into Yip's life. Since Yip did not program a conscience into Victor, he begins to cause problems. Grandfather uses his special effects tricks to put an end to Victor.

Peck, Richard. *The Great Interactive Dream Machine: Another Adventure in Cyberspace.* New York: Dial Books for Young Readers, 1996. 149 pp. (hc. 0-8037-1989-2) $14.99; (pbk. 0-1403-8264-X) $5.99.

Gr. 4–8. Computer geek Aaron Zimmer creates a computer program that grants the wishes of anyone in the area, including a pampered poodle. Aaron and his best friend Josh travel through cyberspace on an assortment of action-packed adventures that are sure to keep readers turning the pages to find out what happens next.

——. *Lost in Cyberspace.* New York: Dial Books for Young Readers, 1995. 160 pp. (hc. 0-8037-1931-0) $16.99; (pbk. 0-1403-7856-1) $5.99.

Gr. 4–8. Sixth-graders Josh and his best friend, a computer geek named Aaron, travel though time with the aid of Aaron's "cellular reorganization" experiment. A housemaid from the 1920s journeys back to the present with the boys and takes over caring for Josh, his sister, and his mother. In their travels, they uncover secrets about the private school they both attend. The young characters are believable and the action is fast paced.

Waugh, Sylvia. *Space Race.* New York: Delacorte, 2000. 241 pp. (hc. 0-385-37266-8) $15.95; (pbk. 0-4404-1714-7) $4.99.

Gr. 4–8. Sent to Earth from the planet Ormingat, Thomas and his father gathered data on the behaviors of earthlings for five years. Eleven-year-old Thomas must now decide whether to return with his father to a planet he does not remember or stay on Earth with their kindly neighbor.

Gutman, Dan. *The Edison Mystery*. New York: Simon & Schuster Books for Young Readers, 2001. 201 pp. (hc. 0-689-84124-8) $16.00.

Gr. 5–9. From the Qwerty Stevens, Back in Time series. When Qwerty finds Edison's Anytime Anywhere Machine, he connects it to his computer and begins traveling through space and time. His trip to Edison's laboratory in 1879 puts him there just in time to watch as Edison and his colleagues discover a long-lasting filament for their electric light bulb. Vintage photographs and a brief timeline of Edison's life are included in the book.

——. *Stuck in Time with Benjamin Franklin*. New York: Simon & Schuster Books for Young Readers, 2002. 182 pp. (hc. 0-689-84553-7) $16.00.

Gr. 5–9. From the Qwerty Stevens, Back in Time series. With a report on the American Revolution due, Qwerty turns on his computer, downloads a paper from the Internet, and pastes a picture of Benjamin Franklin into the paper. Unfortunately, Edison's Anytime Anywhere Machine is plugged into Qwerty's computer and Benjamin Franklin suddenly appears in Qwerty's bedroom. Franklin's time in the twenty-first century shows readers a personal side of him, one not found in history books.

Sidman, Joyce. *Eureka! Poems about Inventors*. Illus. by K. Bennett Chavez. Brookfield, Conn.: The Millbrook Press, 2002. 48 pp. (hc. 0-7613-1665-5) $23.66.

Gr. 5–12. Lyrical, informative poems enhanced by colorful, captivating illustrations tell the stories of famous inventors and their inventions. The poems are separated into four sections: the tapestry of the past, the age of invention, a light interlude, and dawn of the modern age. At the end of each section, there are brief paragraphs about the inventors.

Skinner, David. *The Wrecker*. New York: Simon & Schuster, 1995. 106 pp. (hc. 0-671-79771-9) $14.00.

Gr. 6–9. Theo decides to seek revenge on a bully, Jeffrey, and declares that Michael should be his ally. Theo has the unique ability to walk through a junkyard and know what pieces of junk are calling him. In his hands, the pieces of junk come together to form wondrous mechanical things. Together Theo and Michael channel Theo's unusual talent to create a mechanical wrecking ball to take care of Jeffrey.

Bradbury, Ray. *The Martian Chronicles*, 40th anniversary ed. New
York: Doubleday, 1990. 204 pp. (hc. 0-385-05060-7) $19.95.
Gr. 7–12. Bradbury has written a foreward for this anniversary edition
that describes how his Martian stories became a novel. Travelers from
Earth arrive on Mars to conquer the planet and instead are conquered
by the planet. Interesting parallels can be drawn between the colonization of America and the colonization of Mars, including the devastating
effects of smallpox.

Cooney, Caroline. *Both Sides of Time.* New York: Laurel-Leaf Books,
1995. 240 pp. (pbk. 0-4402-1932-9) $5.50.
Gr. 7–12. Upset by her parents' marital discord, teenager Annie Lockwood laments the lack of romance in her own relationship. While
looking at the old Stratton mansion about to be razed, Annie "falls back
in time" 100 years—into 1895 where she meets Hiram "Strat" Stratton.
Strat is to be the heir to the Stratton mansion and fortune if he will
marry his shy and innocent cousin. Of course Annie and Strat fall
madly in love, creating problems in 1895 AND 1995. This is the first in
a quartet of books about Annie Lockwood and time travel.

——. *For All Time.* New York: Laurel-Leaf Books, 2003. 272 pp. (pbk.
0-4402-2931-6) $4.99.
Gr. 7–12. In this latest book in the time-travel series, Cooney shows
Annie trying to reach Egypt in 1899 where she believes Strat to be
photographing an archeological dig. But Time overshoots the mark and
takes Annie back to Ancient Egypt where she is walled into a tomb as a
sacrifice. Strat has his own problems when his evil father shows up
accusing Strat of being a tomb robber. Whether the "starcrossed" lovers
are united is the impetus for this fourth book.

——. *Out of Time.* New York: Laurel-Leaf Books, 1997. 240 pp. (pbk.
0-4402-1933-7) $4.99.
Gr. 7–12. This sequel to *Both Sides of Time* takes Annie back to the
1890s, a little later than her first visit. She finds that Strat has been
shackled inside an insane asylum for trying to convince people that she
had come from the future. Annie attempts to rescue Strat, to the ire of
villain Walker Walkley. She also finds Strat's cousin Harriet dying of
tuberculosis in an Adirondack sanitarium.

——. *A Prisoner of Time.* New York: Laurel-Leaf Books, 1997. 224 pp.
(pbk. 0-4402-2019-X) $4.99.

Gr. 7–12. This story is not really about Annie Lockwood and Strat. It is about Strat's sister, Devonny, and Annie's brother, Tod. The things that happen in this edition of the time-travel books have an effect on Annie and Strat's romance but are not directly connected.

Cross, Gillian. *New World.* New York: Holiday House, 1994. 171 pp. (hc. 0-8234-1166-4) $15.95.
Gr. 7–12. Delighted to have been chosen "at random" to test a virtual reality game for a computer games company, Miriam is thrilled to give up three hours a week for three weeks with a teenaged boy who was also chosen. The game is exciting for both of them until a really strange thing happens. Stuart is deathly afraid of spiders—a fact known only to his family and himself. Miriam has had a recurring nightmare all her life—a fact known only to her father and herself. Suddenly the "game" is infiltrated with spiders and the sequences in her dream! The ending is extremely satisfying to young adult readers.

Dickinson, Peter. *The Kin.* Illus. by Ian Andrew. New York: G. P. Put-nam's Sons, 1998. 628 pp. (hc. 0-399-24022-5) $24.99.
Gr. 7–12. It is Africa, 200,000 years ago. People are divided up into clans called kins. This book explores the lives of the Moonhawk Kin through the eyes of four young people: Suth, Noli, Ko, and Mana. They have the ability to speak, but they are learning almost everything else. They have fire but must develop a way to carry it with them on their nomadic wanderings in search of food. One of the youngsters discovers a way to weave reeds into mats for carrying things. They shape hard stones into "cutters," which they use for a myriad of things. Almost every day brings new discoveries.

Verne, Jules. *From Earth to the Moon.* New York: Dell, 1993. 185 pp. (pbk. 0-553-21420-9) $5.95.
Gr. 7–12. Verne wrote this story while the Civil War was still raging in America. The setting for the story is at the end of the Civil War in Bal-timore. The members of a gun club decide to shoot a manned projectile to the moon. Hence the premise for this imaginative, prophetic novel about man's race to the moon.

Wells, H. G. *The Time Machine.* New York: Dover Publications, 1995. 76 pp. (pbk. 0-486-28472-2) $3.95.
Gr. 7–12. Originally published in 1895, this short novel has become a well-read classic. The Time Traveler recounts his travels to a group of

Victorian gentlemen. He has traveled hundreds of thousands of years into the future and brought back a glimpse of what is to come.

——. *War of the Worlds*. Rutland, Vt.: J. M. Dent, 1993. 188 pp. (pbk. 0-460-87303-2) $3.95.
Gr. 7–12. When their own planet becomes uninhabitable, the Martians invade Earth with their lethal heat rays. Issues raised by this story are still relevant today, such as whether or not the harmful effects of technology outweigh the benefits. Based on the 1924 Atlantic edition this paperback includes a chronology of Well's life, an introduction, remarks from critics, suggestions for further reading, and a text summary.

Lowry, Lois. *The Giver*. New York: Houghton Mifflin, 1993. 192 pp. (hc. 0-3956-4566-2) $16.00.
Gr. 8–12. In the world of the future, there is no crime, no poverty, no unemployment, no illness and every family is a happy one. Jonas, twelve years old, is chosen to be the Receiver of Memories for the community. The Elders and an old man known as the Giver provide information that ultimately disturbs Jonas as he learns the truth about his utopian world. He struggles with its hypocrisy and wonders at the cost of this pain-free society. This is a Newbery Medal book.

Halperin, James L. *The First Immortal: A Novel of the Future*. New York: Del Rey, 1998. 342 pp. (hc. 0-345-42092-6) $24.95.
Gr. 9–12. This novel looks intensely at the sciences of molecular technology, artificial intelligence, nano-neuroscience, cloning, and cryonics. This fascinating book takes the lives of Dr. Benjamin Smith, his family, and friends from his childhood to 2104. Regardless of the opinions of the reader, the book should provoke interesting discussions.

Chapter 5

Everyday Life

The clock, the sewing machine, the light bulb, the camera, and the toaster are examples of everyday objects whose invention and evolution are chronicled in the books below. These are objects that we do not think about, but that we depend on to get us places on time, to sew the clothes we wear, to enable us to see in the dark, to record what happens in our lives, and to make our breakfast toast.

Wells, Robert E. *How Do You Know What Time It Is?* Morton Grove, Ill.: Albert Whitman, 2002. 32 pp. (hc. 0-8075-7939-4) $14.95; (pbk. 0- 8075-7940-8) $6.95.
Gr. 2–4. Cartoon characters in pen and acrylics with hand-lettered text make this a fun book to read. Wells describes the history of telling time from stick clocks to quartz to atomic clocks.

Maestro, Betsy. *The Story of Clocks and Calendars: Marking a Millennium.* Illus. by Giulio Maestro. New York: Lothrop, Lee and Shepard, 1999. 48 pp. (hc. 0-688-14548-5) $15.95.
Gr. 2–5. From calendars to mark the days, weeks, and years to clocks to mark the seconds, minutes, and hours this is an engaging examination of how humans have measured time. The book closes with additional information about measuring time, a glossary, and an index.

Carlson, Laurie. *Queen of Inventions: How the Sewing Machine Changed the World.* Brookfield, Conn.: The Millbrook Press, 2003. 32 pp. (hc. 0-7613-2706-1) $22.90.
Gr. 3–6. Black and white photographs, reproductions, and succinct text explain how the sewing machine saved time and revolutionized sewing

by making possible the mass production of clothing and other items. Isaac Singer was not only an inventor, he was also a shrewd business-person. He allowed people to make down payments on sewing machines and as they earned money sewing, they were able to make the remaining payments. On the last page, there are books for further reading, a bibliography, and websites.

Pollard, Michael. *The Light Bulb and How It Changed the World.* New York: Facts On File, 1995. 46 pp. (hc. 0-8160-3145-2) $23.00.
Gr. 3–6. From the History and Invention series. The inventors and inventions that make the light bulb glow are described in brief paragraphs, pictures, reproductions, and diagrams. Throughout the book readers learn how the light bulb and electricity influenced daily life. A glossary, an index, and books for further reading conclude the book.

Dolan, Graham. *The Greenwich Guide to Time and the Millennium.* Illus. by Jeff Edwards. Chicago: Heinemann Library, 1999. 48 pp. (hc. 1-57572-802-8) $12.00.
Gr. 4–8. This brief overview of time includes information on sundials, longitude, clock making, calendars, and time zones. John Harrison's chronometer for measuring longitude is given a two-page spread. Photographs, drawings, diagrams, maps, charts, a glossary, more books to read, and an index are provided.

Duffy, Trent. *The Clock.* New York: Atheneum Books for Young Readers, 2000. 80 pp. (hc. 0-689-82814-4) $17.95.
Gr. 4–8. From the Turning Point Inventions series. From Stonehenge to atomic clocks, this is a fascinating look at the innovations in time-keeping including Harrison's chronometer for measuring longitude. Books for further reading and an index conclude the book.

Wilsdon, Christina. *Everyday Things: An A–Z Guide.* New York: Franklin Watts, 2001. 128 pp. (hc. 0-531-11792-8) $32.00; (pbk. 0-531-15451-3) $19.95.
Gr. 4–8. From the Watts Reference series. Discover the origins of everyday objects such as fabric, ice cream, microwave ovens, trading cards, and Velcro in this well-written book. Color photographs; sidebars; bold, blue subheadings; and clear, concise entries help make the information easily accessible to readers. A bibliography and an index conclude the book.

Tchudi, Stephen. *Lock and Key: The Secrets of Locking Things Up, In, and Out.* New York: Charles Scribner's Sons, 1993. 128 pp. (hc. 0-684-19363-9) $14.95.
Gr. 4–12. Readers travel back in time to discover the earliest of locks and end their journey with biometric systems that read fingerprints and eyes. There is information about picking locks with words of caution to leave this activity to bonded locksmiths. Black and white illustrations and photographs accompany the text. References and an index conclude the book.

Wallace, Joseph. *The Camera.* New York: Atheneum Books for Young Readers, 2000. 79 pp. (hc. 0-689-82813-6) $17.95.
Gr. 4–12. From the Turning Point Inventions series. From the camera obscura to the digital camera, Wallace presents a concise history of the development of the camera. Photographs, reproductions, and diagrams portray the development of the camera and show how it works. A list of books for further reading and an index are included.

——. *The Lightbulb.* New York: Atheneum Books for Young Readers, 1999. 79 pp. (hc. 0-689-82816-0) $17.95.
Gr. 4–12. From the Turning Point Inventions series. This well-written text briefly describes the invention of the light bulb, portrays Edison's life and work, and ends with information on the future of light. Photographs, illustrations, and diagrams enhance the text and help readers understand the evolution and inner workings of the light bulb. A list of books for further reading and an index are included.

Dowswell, Paul. *Everyday Life.* Chicago: Heinemann Library, 2001. 48 pp. (hc. 1-58810-232-1) $25.64.
Gr. 5–8. From the Great Inventions series. This is a captivating examination of innovations that we use every day. The book contains interesting facts about the innovations, photographs, "how it works" diagrams, and mini-biographies of inventors. It is useful as a starting point for more in-depth explorations of the innovations. The book concludes with a timeline, a glossary, more books to read, and an index.

Rubin, Susan Goldman. *Toilets, Toasters, and Telephones.* Illus. by Elsa Warnick. San Diego: Harcourt Brace, 1998. 132 pp. (hc. 0-15-201421-7) $20.00.
Gr. 5–8. Examined in this thought-provoking book are bathroom fixtures, kitchen appliances, appliances for cleaning, telephones, typewrit-

ers, and other everyday objects. Text accompanied by black and white photographs chronicles the invention and progression of these household items through the ages as they changed shape and form. The book introduces students to the importance of industrial design. An afterword, a bibliography, and an index conclude the book.

Levy, Joel. *Really Useful: The Origins of Everyday Things.* Willowdale, Ont.: Firefly Books, 2002. 240 pp. (hc. 1-55297-623-8) $39.95; (pbk. 1-55297-622-X) $19.96.

Gr. 8–12. Non-stick pans, ATM machines, clock radios, and self-adhesive tape are things we use on a daily basis and we do not think much about them. Levy sets out to change all that as he shares the origins of everyday things we take for granted. Glossy color photographs grab the readers' attention and the entries are short and entertaining. Resources for learning more and an index conclude the book.

Friedel, Robert. *Zipper: An Exploration in Novelty.* New York: W.W. Norton, 1994. 288 pp. (hc. 0-393-03599-9) $23.00.

Gr. 10–12. From Whitcomb Judson's patent for a "hookless fastener" for shoes comes the zipper, something people use every day, but do not think about. The evolution from "hookless fastener" to zipper is filled with interesting characters, patent disputes, fashion vagaries, and marketing problems. Included in the book are photographs, technical diagrams, notes, and an index.

Lindsay, David. *House of Invention: The Secret Life of Everyday Products.* New York: The Lyons Press, 2000. 179 pp. (hc. 1-55821-740-1) $22.95.

Gr. 10–12. Every room in your house contains inventions each with their own interesting story. Lindsay takes readers along as he tells the absorbing stories behind inventions found in the bathroom, the kitchen, the foyer, the office, the garage, the family room, and the bedroom. A bibliography concludes the book.

Chapter 6

Industry, Energy, and the Environment

Nonfiction books in this chapter include discoveries and inventions that have created industries, harnessed energy, and impacted the environment. Construction projects, machines, the Industrial Revolution, lasers, superconductors, and alternative energy sources are some of the topics found in the books annotated below.

Hoban, Tana. *Construction Zone*. New York: Greenwillow Books, 1997. 32 pp. (hc. 0-688-12285-X) $15.95.
Gr. P–1. Thirteen color photographs of different pieces of construction equipment introduce young readers to the construction process. On the left side of each two-page spread is a photograph of the machine and on the right side is a close up of part of the machine. At the end of the book are thumbnail shots of the machines and brief descriptions of what they do.

Landau, Elaine. *Bridges*. New York: Children's Press, 2001. 48 pp. (hc. 0-516-22182-5) $23.50; (pbk. 0-516-27313-2) $6.95.
Gr. P–2. From a True Book series. Large print and color photographs introduce readers to different kinds of bridges and how they are constructed. Resources for learning more, a glossary, and an index conclude the book.

——. *Canals*. New York: Children's Press, 2001. 48 pp. (hc. 0-516-22183-3) $23.50; (pbk. 0-516-27314-0) $6.95.

Gr. P–2. From a True Book series. The book describes how canals are built and how locks work. Then it provides specific information on the construction and history of the Erie Canal, Suez Canal, and Panama Canal. Resources for learning more, a glossary, and an index conclude the book.

——. *Skyscrapers*. New York: Children's Press, 2001. 48 pp. (hc. 0-516-22184-1) $23.50; (pbk. 0-516-27324-8) $6.95.
Gr. P–2. From a True Book series. This is a brief history of skyscrapers told in a large font with accompanying color photographs. The engineer and architect William Le Baron Jenny designed the first skyscraper, the Home Insurance Company Building in Chicago. Resources for learning more, a glossary, and an index conclude the book.

——. *Tunnels*. New York: Children's Press, 2001. 48 pp. (hc. 0-516-22185-X) $23.50; (pbk. 0-516-27325-6) $6.95.
Gr. P–2. From a True Book series. Tunnels are dug under the ground, through the mountains, and under the water. This easy to read book provides basic facts about tunnel construction and looks at tunnels around the world including the Chunnel stretching between England and France. Resources for learning more, a glossary, and an index conclude the book.

Hill, Lee Sullivan. *Dams Give Us Power*. Minneapolis: Carolrhoda Books, 1997. 32 pp. (hc. 1-57505-023-4) $14.60.
Gr. K–2. Bright, colorful photographs and short paragraphs introduce children to dams, describe how they are used, and explain how they are constructed. A photo index contains thumbnail sketches and information about each of the dams mentioned in the book.

Sirimarco, Elizabeth. *At the Construction Site*. Chanhassen, Minn.: The Child's World, 2000. 32 pp. (hc. 1-56766-573-X) $22.79.
Gr. K–2. From the Field Trips series. Large, colorful photographs, large print, and a cartoon character guide with a parrot perched on his shoulder introduce young children to a construction site. The glossary is for adults to use to explain the terms in bold print throughout the book. An index is provided.

Thomas, Mark. *The Akashi Kaikyo Bridge: World's Longest Bridge*. New York: PowerKids Press, 2002. 24 pp. (hc. 0-8239-5990-2) $17.25.

Gr. 2–3. From the Record-Breaking Structures series. Color photographs and short sentences explain how the bridge was constructed. A note at the end of the book states that this Reading Power book is easy to read with high content for struggling readers. A glossary, resources, and an index are provided.

Halperin, Wendy Anderson. *Once Upon a Company*. New York: Orchard Books, 1998. 32 pp. (hc. 0-531-30089-7) $16.95.
Gr. 2–5. One cold, wet November day Halperin suggested that her bored children make wreaths to sell at Christmas. Their grandfather suggested they save any money they earned for college; hence, the College Fund Wreath Company was formed. In the summer, they formed the Peanut Butter and Jelly Company and sold sandwiches. Each year their businesses continue to grow as does their college fund. A glossary ends the book.

Boekhoff, P. M., and Stuart A. Kallen. *Lasers*. San Diego: KidHaven, 2002. 47 pp. (hc. 0-7377-0944-8) $23.70.
Gr. 3–5. From the KidHaven Press Science Library series. This book provides a brief overview of the development of lasers, how they are used today, and possible future uses. Color photographs, diagrams, and large fonts entice readers to explore this fascinating subject. A glossary, resources for learning more, and an index are included.

Collins, Mary. *The Industrial Revolution*. New York: Children's Press, 2000. 32 pp. (hc. 0-516-21596-5) $21.00.
Gr. 3–5. From the Cornerstones of Freedom series. Beginning in Great Britain the Industrial Revolution changed the way people lived and worked. Collins describes the impact of the Revolution on America and introduces John Muir. He championed the protection of natural resources at a time when they were being consumed by industries with little thought to conservation or preservation. A glossary, a timeline, and an index conclude the book.

Jones, George. *My First Book of How Things Are Made: Crayons, Jeans, Peanut Butter, Guitars, and More*. New York: Scholastic, 1995. 64 pp. (hc. 0-590-48004-9) $12.95.
Gr. 3–5. Children reading this book discover how some of their favorite things are made. Items included are crayons, peanut butter, grape jelly, footballs, orange juice, blue jeans, guitars, and books. The steps in the

manufacturing process are numbered and described in brief paragraphs, which are accompanied by clear, color photographs.

Brown, David J. *The Random House Book of How Things Were Built.* New York: Random House, 1992. 140 pp. (hc. 0-679-82044-2) $15.00.
Gr. 3–6. Pyramids, Stonehenge, Machu Picchu, and the Taj Mahal are a few of the buildings whose engineering and architectural designs are examined in this illustrated book. The discussion on each building is limited to two-page spreads, which contain just enough information for browsing and enticing readers to learn more about these structures. A glossary and an index conclude the book.

Curlee, Lynn. *Brooklyn Bridge.* New York: Atheneum Books for Young Readers, 2001. 36 pp. (hc. 0-689-83183-8) $18.00.
Gr. 3–6. Diagrams of the Brooklyn Bridge, cross sections, a map, and acrylic paintings enhance the text and help readers understand the engineering and construction details. Bridge specifications, a timeline, and a bibliography conclude the book.

Stone, Tanya Lee. *America's Top Ten Construction Wonders.* Woodbridge, Conn.: Blackbirch Press, 1998. 24 pp. (hc. 1-56711-195-5) $17.95.
Gr. 3–6. From America's Top Ten series. The very brief introductions with accompanying full-page color photographs can be used as the starting point for more a more in-depth study of these construction wonders. Included in the book are the Alaska Pipeline, Epcot Center, the Erie Canal, the Gateway Arch, the Grand Coulee Dam, the Hubble Space Telescope, the Lincoln Tunnel, the Louisiana Superdome, the Sears Tower, and the Seattle Space Needle. At the end of the book is a list of other American construction wonders, a glossary, books for further reading, online resources, and an index.

Wilkinson, Philip. *Super Structures.* New York: Dorling Kindersley, 1996. 44 pp. (hc. 0-7894-1011-7) $15.95.
Gr. 3–6. Diagrams, cutaway views, three-dimensional models, and photographs show how super structures are constructed. Skyscrapers, tunnels, bridges, nuclear reactors, and oilrigs are just a few of the structures found in this absorbing book. A glossary and an index are included.

Graham, Ian. *Solar Power*. Austin, Tex.: Raintree Steck-Vaughn, 1999. 48 pp. (hc. 0-8172-5362-9) $27.12.
Gr. 4–6. From the Energy Forever Series. Sunlight provides the earth with a renewable source of energy. This book explains the technology needed to harness solar power and describes the future of solar power. Photographs, diagrams, reproductions, and sidebars focus readers' attention on important concepts presented in the text. A glossary, a further information section, and an index conclude the book.

——. *Water Power*. Austin, Tex.: Raintree Steck-Vaughn, 1999. 48 pp. (hc. 0-8172-5363-7) $27.12.
Gr. 4–6. From the Energy Forever Series. Water is a renewable energy source and this book describes how its power has been harnessed by watermills and huge dams. Photographs, diagrams, reproductions, and sidebars focus readers' attention on important concepts. A glossary, a further information section, and an index conclude the book.

——. *Wind Power*. Austin, Tex.: Raintree Steck-Vaughn, 1999. 48 pp. (hc. 0-8172-5364-5) $27.12.
Gr. 4–6. From the Energy Forever Series. Windmills and wind turbines harness the wind, providing a free source of energy that generates electricity and pumps water. Past, current, and future uses for wind power are described in this book. Photographs, diagrams, reproductions, and sidebars focus readers' attention on important concepts presented in the text. A glossary, a further information section, and an index conclude the book.

Hoare, Stephen. *The World of Caves, Mines, and Tunnels*. Illus. by Bruce Hogarth. New York: Peter Bedrick Books, 1999. 45 pp. (hc. 0-87226-294-4) $19.95.
Gr. 4–6. Colorful illustrations, cutaway views, and brief descriptive text explore the world beneath us. The book explains how people live, work, and travel underground and examines some of the technology that makes this possible. A glossary and an index conclude the book.

Kent, Peter. *Great Buildings: Stories of the Past*. New York: Oxford University Press, 2001. 47 pp. (hc. 0-19-521846-9) $18.95.
Gr. 4–6. Travel back in time and around the world to examine the construction techniques used to create nine different structures including the Eiffel Tower, the Great Wall of China, and the Brooklyn Bridge.

Cutaway diagrams, illustrations, and maps help explain the text. An index concludes the book.

Wilcox, Charlotte. *Powerhouse: Inside a Nuclear Power Plant*. Illus. by Jerry Boucher. Minneapolis: Carolrhoda Books, 1996. 48 pp. (hc. 0-87614-945-X) $22.60; (pbk. 0-87614-979-4) $15.88.
Gr. 4–6. Wilcox takes readers inside a nuclear power plant where they learn how nuclear power produces electricity. Photographs, diagrams, and separate text boxes contain additional information and explanations. A glossary and an index conclude the book.

Aaseng, Nathan. *The Unsung Heroes: Unheralded People Who Invented Famous Products*. Minneapolis: Lerner, 1989. 80 pp. (hc. 0-8225-0676-0) $18.95.
Gr. 4–8. The book briefly introduces the people behind brand name products. The products are Band-Aids, Coca-Cola, Dunlop tires, General Motors vehicles, Hoover vacuum cleaners, bingo, Superman comics, Hummel figurines, and McDonalds fast food. Black and white photographs, a bibliography, and an index are included in the book.

Adkins, Jan. *Bridges: From My Side to Yours*. Brookfield, Conn.: Roaring Brook Press, 2002. 96 pp. (hc. 0-7613-1542-X) $23.90.
Gr. 4–8. From stepping stones to arches to cables, this is a concise history of bridge building accompanied by detailed black and white drawings. The author explores the design, architecture, and engineering aspects of constructing bridges from the simple to the complex. A glossary and an index conclude the book.

Bial, Raymond. *The Canals*. Tarrytown, N.Y.: Marshall Cavendish, 2002. 55 pp. (hc. 0-7614-1336-7) $25.64.
Gr. 4–8. From the Building America series. While most recognize George Washington as the father of our country, they do not realize that he is also considered to be the father of canal building in America. This and other interesting pieces of information can be gleaned from this overview of canal building in America. A glossary, resources for further learning, a bibliography, and an index conclude the book.

——. *The Farms*. Tarrytown, N.Y.: Marshall Cavendish, 2002. 55 pp. (hc. 0-7614-1332-4) $25.64.
Gr. 4–8. From the Building America series. Step back into America's rural past beginning in Colonial America and see how houses, roads,

fences, barns, and outbuildings evolved. Photographs, reproductions, and drawings capture bygone structures. A glossary, resources for further learning, a bibliography, and an index conclude the book.

——. *The Forts*. Tarrytown, N.Y.: Marshall Cavendish, 2002. 55 pp. (hc. 0-7614-1334-0) $25.64.
Gr. 4–8. From the Building America series. This history of American forts shows how their construction changed over the years. Photographs and reproductions depict the changes in the forts. A glossary, resources for further learning, a bibliography, and an index conclude the book.

——. *The Houses*. Tarrytown, N.Y.: Marshall Cavendish, 2002. 55 pp. (hc. 0-7614-1335-9) $25.64.
Gr. 4–8. From the Building America series. Colonial houses, log cabins, dugouts, soddies, and other houses from across the American landscape show how houses were constructed of available materials and how they changed over time. A glossary, resources for further learning, a bibliography, and an index conclude the book.

——. *The Mills*. Tarrytown, N.Y.: Marshall Cavendish, 2002. 55 pp. (hc. 0-7614-1333-2) $25.64.
Gr. 4–8. From the Building America series. Sawmills, windmills, gristmills, and watermills are described in clear text accompanied by color photographs and illustrations. Diagrams help explain how the different mills operate. A glossary, resources for further learning, a bibliography, and an index conclude the book.

Doherty, Katherine M. *The Gateway Arch*. Tarrytown, N.Y.: Marshall Cavendish, 2002. 55 pp. (hc. 0-56711-105-X) $25.64.
Gr. 4–8. From the Building America series. Spectacular photographs taken from the top of the arch in St. Louis have the power to make the susceptible reader dizzy. Details of the design and the construction of the arch demonstrate that this is truly an architectural feat of enormous proportions. A glossary, a chronology, books for further reading, source notes, and an index are included.

Goodall, Jane. *The Chimpanzees I Love: Saving Their World and Ours*. New York: Scholastic Press, 2001. 80 pp. (hc. 0-439-21310-X) $17.95.
In absorbing narrative Goodall describes her observations and discoveries about chimpanzees. Compelling color photographs help readers

understand the importance of protecting the chimpanzees. The book concludes with a facts and resources section that contains information on chimpanzees, a primate family tree, a map of Africa, contact information for Roots and Shoots, Goodall's environmental education program, and resources for learning more.

Gourley, Catherine. *Good Girl Work: Factories, Sweatshops, and How Women Changed Their Role in the American Workforce.* Brookfield, Conn.: The Millbrook Press, 1999. 96 pp. (hc. 0-7613-0951-9) $26.90.
Gr. 4–8. With the coming of the Industrial Revolution in the nineteenth and twentieth centuries, women began working in factories and sweatshops. They worked long hours for low pay at often dangerous, monotonous jobs. This book chronicles their work and their efforts to make changes. Photographs and quotes from the girls and women help readers make personal connections between the character's lives and their own. An epilogue, notes, and an index conclude the book.

Mann, Elizabeth. *Hoover Dam.* Illus. by Alan Witschonke. New York: Mikaya Press, 2001. 48 pp. (hc. 1-931414-02-5) $19.95.
Gr. 4–8. From the Wonders of the World series. This well-organized book contains a great deal of information about the construction of the Hoover Dam, including quotes from those who worked on the dam. Facts about the dam, a list of those who died in the construction of the dam, and an index are included.

——. *The Panama Canal.* Illus. by Fernando Rangel. New York: Mikaya Press, 1998. 48 pp. (hc. 0-9650493-4-5) $19.95.
Gr. 4–8. From the Wonders of the World Book series. Descriptive, engaging narrative, paintings, and reproductions tell the history of the construction of the Panama Canal. A foldout diagram of the canal locks, a map of the canal, an author's note, a list of facts about the canal, and an index are included.

Rubin, Susan Goldman. *There Goes the Neighborhood: Ten Buildings People Loved to Hate.* New York: Holiday House, 2001. 96 pp. (hc. 0-8234-1435-3) $19.95.
Gr. 4–8. Now world-renowned, these ten buildings and their architects described were ridiculed and despised by the community. The Eiffel Tower and the Washington Monument are two of the controversial buildings. A glossary, notes on some of the architects, a selected bibli-

ography, resources for learning more, and an index are located at the end of the book.

McCormick, Anita Louise. *The Industrial Revolution in American History.* Springfield, N.J.: Enslow, 1998. 128 pp. (hc. 0-89490-985-1) $20.95.
Gr. 4–10. From the In American History series. Beginning with the Industrial Revolution in England, to the spread of the Revolution to America, and its decline in the twentieth century this is a concise overview of an exciting, turbulent time in history. The factory system, transportation, communication, and big business are all discussed in separate chapters. Black and white photographs, a timeline, chapter notes, resources for further reading, and an index are included.

Aaseng, Nathan. *Construction: Building the Impossible.* Minneapolis: The Oliver Press, 2000. 144 pp. (hc. 1-881508-59-5) $20.50.
Gr. 5–8. From the Innovators series. Construction projects featured in the book are the Step Pyramid, the Thames Tunnel, the Brooklyn Bridge, the Eiffel Tower, the Panama Canal, the Hoover Dam, and the Empire State Building. Each construction project has its own chapter, which contains a description of the project, a profile of the builder, and black and white photographs. A list of the tallest buildings, a list of the longest bridges, a glossary, a bibliography, and an index conclude the book.

——. *The Fortunate Fortunes.* Minneapolis: Lerner, 1989. 76 pp. (hc. 1-8225-0678-5) $15.95.
Gr. 5–8. Hard work, persistence, imagination, and sometimes a little luck are required for an invention to become financially viable. These are the success stories behind Hires Root Beer, Johnson Wax, Wrigley's Gum, Kellogg's Corn Flakes, Baby Ruth, Kleenex Tissue, Owens-Corning, and Encyclopaedia Britannica. Black and white photographs, a bibliography, and an index are included.

——. *The Problem Solvers.* Minneapolis: Lerner, 1989. 80 pp. (hc. 1-8225-0675-0) $15.95.
Gr. 5–8. Some people turn problems into opportunities to produce useful products and then turn those products into companies. This book contains the stories behind these companies: John Deere, Prudential, Kitchen-Aid Dishwasher, Westinghouse, Evinrude, Gerber, Polaroid,

Jacuzzi, and AstroTurf. Black and white photographs, a bibliography, and an index are included.

——. *The Rejects*. Minneapolis: Lerner, 1989. 80 pp. (hc. 1-8225-0677-7) $15.95.
Gr. 5–8. These are the stories of familiar products that were initially rejected and deemed failures. Brief, out of the ordinary stories about these products are contained in the book: graham crackers, Borden condensed milk, *Reader's Digest*, Birdseye frozen foods, Monopoly, Xerox, the Lear jet, and Orville Redenbacher popcorn. Black and white photographs, a bibliography, and an index are included.

Levy, Janey. *The Erie Canal: A Primary Source History of the Canal That Changed America*. New York: Rosen Publishing Group, 2003. 63 pp. (hc. 0-8239-3680-5) $29.25.
Gr. 5–8. From the Primary Sources in American History series. More than a history of the Erie Canal, this book also demonstrates the use of primary source documents in research and writing. Illustrations with detailed captions contain additional information that add to the text and provide information on the construction of the canal. A timeline, primary source transcriptions, a glossary, resources for more information, a bibliography, an index, and a primary source list conclude the book.

Morgan, Nina. *Lasers*. Austin, Tex.: Raintree Steck-Vaughn, 1997. 48 pp. (hc. 0-8172-4812-9) $27.12.
Gr. 5–8. From the Twentieth Century Inventions series. The word "laser" is actually an acronym for "light amplification by simulated emission of radiation." From there, Morgan explains how lasers work and how they are used in communications, war, science, industry, and health. Large color photographs and diagrams draw readers into the book and enhance the descriptions provided in the text. A date chart, a glossary, a list of books to find out more, and an index are located at the end of the book.

Pringle, Laurence. *Global Warming: The Threat of Earth's Changing Climate*. New York: SeaStar Books, 2001. 48 pp. (hc. 1-58717-009-4) $16.95.
Gr. 5–8. This book is a powerful examination of the problem of global warming that discusses the consequences and possible solutions including the Kyoto Protocol. This protocol is actually a plan whereby thirty-eight industrialized nations have agreed to cut their harmful

emissions by 2012. Colorful photographs, diagrams, and charts combined with vivid writing grab readers' attention. A glossary, resources for further reading, and an index are included.

Severance, John B. *Skyscrapers: How America Grew Up.* New York: Holiday House, 2000. 82 pp. (hc. 0-8234-1492-2) $18.95.
Gr. 5–8. This book chronicles the development of skyscrapers and the changes they have made in our lives. Innovations that made skyscrapers possible such as steel, elevators, diesel earthmovers, colossal cranes, and modern plumbing are discussed at the beginning of the book. Once the construction challenges of building a skyscraper were overcome, other problems surfaced, such as their swaying in high winds and huge glass windows popping out. Notes, a bibliography, and an index are included.

Vanderwarker, Peter. *The Big Dig: Reshaping an American City.* Boston: Little, Brown, 2001. 56 pp. (hc. 0-316-60598-0) $17.95.
Gr. 5–8. Boston's Central Artery/Tunnel Project, The Big Dig, encompasses new roads, bridges, and tunnels to alleviate the traffic congestion in the city. Photographs, maps, diagrams, and informative text explain the massive undertaking that will involve twenty years of construction. A brief glossary is found at the end of the book.

Cefrey, Holly. *What If the Hole In the Ozone Layer Grows Larger?* New York: Children's Press, 2002. 49 pp. (hc. 0-516-23913-9) $20.00.
Gr. 5–9. From the What If series. This book provides a basic introduction to the environmental problems attributed to the hole in the Earth's ozone layer. Photographs, diagrams, and sidebars contain additional information to help readers understand this environmental problem. A glossary, resources for learning more, and an index are included.

Daley, Michael J. *Nuclear Power: Promise or Peril?* Minneapolis: Lerner, 1997. 143 pp. (hc. 0-8225-2611-5) $25.25.
Gr. 5–9. From the Pro/Con series. Daley begins the debate by describing a nuclear power plant, Vermont Yankee that safely produces an abundance of electricity. Then, he describes the Chernobyl Nuclear Power Station disaster. He explains how nuclear power works and gives a brief history of energy. This provides the basic information students need to understand the pros and cons of nuclear energy presented

in the remainder of the book. Resources to contact, endnotes, a glossary, a bibliography, and an index are provided.

Darling, David. *Micromachines and Nanotechnology: The Amazing New World of the Ultrasmall.* Parsippany, N.J.: Dillon, 1995. 64 pp. (pbk. 0-382-24953-4) $11.00.
Gr. 5–9. From the Beyond 2000 series. Explore the possibilities of tiny machines and look into the future when nanotechnology maybe used to cleanse blood streams, disintegrate hazardous waste, or digest garbage. This informative book is a great starting place to have students thinking about other uses for micromachines. Diagrams, photographs, sidebars, glossary, books for further reading, and an index are included.

Parker, Steve. *Fuels for the Future.* Austin, Tex.: Raintree Steck-Vaughn, 1998. 48 pp. (hc. 0-8172-4934-6) $27.12.
Gr. 6–8. Parker describes a variety of fuels used around the world such as gasoline, animal dung, and steam. He discusses problems created by burning fuels and then describes steps to take to reduce our use of fuel and our dependency on fossil fuels. Photographs, charts, diagrams, and maps help explain concepts presented in the text. A glossary, resources for further research, and an index complete the book.

Colman, Penny. *Rosie the Riveter: Women Working on the Home Front in World War II.* New York: Crown, 1998. 120 pp. (hc. 0-517-88567-0) $10.99.
Gr. 6–9. This award-winning book chronicles women's movement into the work force during World War II and the profound impact this movement had on American industry and society. Photographs, posters, and advertisements help tell this dramatic story. A select list of women's wartime jobs, facts and figures about women war workers, a chronology, a bibliography, notes, and an index conclude the book.

DuTemple, Lesley A. *The New York Subways.* Minneapolis: Lerner, 2003. 79 pp. (hc. 0-8225-0378-6) $27.93.
Gr. 6–9. From the Great Building Feats series. DuTemple chronicles the evolution of the New York subway system and includes information on the construction techniques, the financing, and the politics. Photographs, diagrams, and sidebars of mini-biographies and interesting facts complement the text. Source notes, a selected bibliography, resources for further reading, and an index are included.

——. *The Panama Canal.* Minneapolis: Lerner, 2003. 95 pp. (hc. 0-8225-0079-5) $27.93.

Gr. 6–9. From the Great Building Feats series. Photographs, diagrams, sidebars of mini-biographies, and sidebars of interesting facts complement the text. A timeline of the Panama Canal, source notes, a selected bibliography, resources for further reading, and an index are included.

Woods, Michael, and Mary B. Woods. *Ancient Machines: From Wedges to Waterwheels.* Minneapolis: Lerner, 2000. 88 pp. (hc. 0-8225-2994-7) $25.26.

Gr. 6–9. From the Ancient Technology series. Descriptions of simple machines from the ancient Middle East, Egypt, China, Greece, and Rome include connections to current technology. Diagrams, maps, and photographs accompany the well-written text. The book concludes with a glossary, a selected bibliography, and an index.

Karwatka, Dennis. *Technology's Past: America's Industrial Revolution and the People Who Delivered the Goods.* Ann Arbor, Mich.: Prakken Publications, 1999. 262 pp. (pbk. 0-911168-91-5) $29.95.

Gr. 6–12. Brief profiles of selected inventors and their inventions are arranged chronologically from the Industrial Revolution to World War II in this concise collection. The entries include photographs and diagrams, concluding with a short list of references and resources. General overviews of computers, television, space flight, and robotics are at the end of the book. An index is included.

Macaulay, David. *Building Big.* Boston: Houghton Mifflin, 2000. 192 pp. (hc. 0-395-96331-1) $30.00.

Gr. 6–12. Examine the construction of bridges, tunnels, dams, domes, and skyscrapers from around the world and over the years. Detailed illustrations, cutaway views, and labeled diagrams accompany the text as Macaulay explains the designing and engineering feats that made the construction possible. This book is a companion to Macaulay's PBS series. At the end of the book is a brief glossary.

——. *Building the Book Cathedral.* Boston: Houghton Mifflin, 1999. 112 pp. (hc. 0-395-92147-3) $29.95.

Gr. 6–12. In this book, Macaulay immerses readers in the process of creating his book *Cathedral,* which is annotated below. The original book *Cathedral* is incorporated into *Building the Book Cathedral* along with additional sketches and notes on why and how changes were made as work on *Cathedral* progressed. Some of these notes help readers understand the importance of perspective, scale, and contrast in drawing. Other notes describe what he thinks he should have done differently when he wrote *Cathedral*. A glossary is provided.

——. *Castle*. Boston: Houghton Mifflin, 1977. 80 pp. (hc. 0-395-25784-0) $18.00.
Gr. 6–12. Travel back to thirteenth-century Wales and watch the construction of a castle and an adjoining town. Detailed drawings and descriptive narrative explain the enormous undertaking. A glossary is located at the back of the book.

——. *Cathedral: The Story of Its Construction*. Boston: Houghton Mifflin, 1974. 80 pp. (hc. 0-395-17513-5) $18.00; (pbk. 0-395-31668-5) $8.95.
Gr. 6–12. Follow the step-by-step construction of this thirteenth-century Gothic cathedral in an imaginary French town. Pen and ink sketches and detailed descriptive narrative are testimony to the extensive research done for this book.

——. *Pyramid*. Boston: Houghton Mifflin, 1975. 80 pp. (hc. 0-395-21406-0) $18.00; (pbk. 0-395-32121-2) $8.95.
Gr. 6–12. Detailed line drawings and concise text show how the ancient Egyptians built their pyramids. This book gives readers an opportunity to explore the engineering and architectural marvels of the pyramids.

Pringle, Laurence. *The Environmental Movement: From Its Roots to the Challenges of a New Century*. New York: HarperCollins, 2000. 144 pp. (hc. 0-688-15626-6) $16.95.
Gr. 6–12. From the Native Americans to Henry David Thoreau to George Perkins Marsh to John Muir to Rachel Carson, the history of the environmental movement has had champions who have understood the importance of preserving our habitat. Their observations and discoveries call attention to environmental concerns. Photographs and cartoons help stress the importance of the environmental movement. The book ends with a look to the future. A list of environmental groups and agencies, a list of government agencies, resources for further reading, and an index are included.

Marshall, Elizabeth. *High–Tech Harvest: A Look at Genetically Engineered Foods*. New York: Franklin Watts, 1999. 144 pp. (hc. 0-531-11434-1) $20.00.
Gr. 7–10. From the Impact Book series. This is a succinct overview of genetically engineered foods that discusses the benefits, ethical considerations, and what the future holds. Black and white photographs and line drawings are included. A glossary, end notes, resources for learning more, and an index conclude the book.

Morgan, Sally. *Alternative Energy Sources*. Chicago: Heinemann Library, 2003. 64 pp. (hc. 1-40340-322-8) $28.50.
Gr. 8–12. From the Science at the Edge series. Fossil fuels, wind, the sun, water, geothermal power, nuclear energy, and bioenergy are the types of fuels discussed in this book. Morgan explains how at some point fossil fuels will run out and that people will have to depend on alternative fuels. She also discusses the future of renewable energy sources. A timeline, a glossary, sources for learning more, and an index are provided.

Collier, Christopher, and James Lincoln Collier. *The Rise of Industry: 1860–1900*. New York: Marshall Cavendish, 2000. 94 pp. (hc. 0-7614-0820-7) $20.95.
Gr. 8–12. From the Drama of American History series. During this time period, America was transformed from rural farm communities to urban industrial cities. Technology was changing the way people lived and worked. The authors chronicle these changes and note their impact on society. A bibliography and an index conclude the book.

Billington, David P. *The Innovators: The Engineering Pioneers Who Made America*. New York: John Wiley and Sons, 1996. 245 pp. (hc. 0-471-14026-0) $24.95.
Gr. 9–12. Billington has written from the perspective of an engineer and presents an intriguing examination of the engineering innovations and the key figures that impacted life in the nineteenth century. The book is divided into two parts: iron, steam, and early industry, 1776–1855 and crossing the continent, 1830–1883. Photographs, reproductions, diagrams, notes, references, and an index are provided.

Bortz, Fred. *Techno–Matter: The Materials behind the Marvels*. Brookfield, Conn.: The Millbrook Press, 2001. 96 pp. (hc. 0-7613-1469-5 $25.90.

Gr. 9–12. Techno-matter makes twenty-first-century inventions possible. The first chapter explains the basic chemistry necessary to understand the other chapters. These materials are explored in the book: conductors, insulators, and semiconductors; digital matter; polymers; metals, alloys, ceramics, and glass; and composites. A glossary, resources for learning more, and an index can be found at the end of the book.

Perlin, John. *From Space to Earth: The Story of Solar Energy*. Ann Arbor, Mich.: Aatec Publications, 1999. 224 pp. (hc. 0-937948-14-4) $32.00.
Gr. 9–12. Photovoltaics provide an electrical supply where it is not feasible to run wires and where other sources of energy are not feasible. From oil platforms to lighthouses to buoys, solar cells provide electric power in remote locations. Black and white photographs and an index are included.

Hecht, Jeff. *City of Light: The Story of Fiber Optics*. New York: Oxford University Press, 1999. 316 pp. (hc. 0-19-510818-3) $29.95.
Gr. 10–12. This is a lively account of the history of fiber optics and how it has impacted various aspects of life including communications and medicine. Appendixes, notes, and an index are provided.

Slack, Charles. *Noble Obsession: Charles Goodyear, Thomas Hancock, and the Race to Unlock the Greatest Industrial Secret of the Nineteenth Century*. New York: Hyperion, 2002. 274 pp. (hc. 0-7868-6789-2) $24.95.
Gr. 10–12. Goodyear's relentless pursuit of a way to stabilize rubber regardless of its temperature landed him in debtors' prison, tied him up in courtroom dramas, and left him penniless. This is a compelling account of the challenges and obstacles faced by one inventor as he attempted to protect his rights to an invention and make it financially viable. Sources, a bibliography, and an index conclude the book.

Tobin, James. *Great Projects: The Epic Story of the Building of America, from the Taming of the Mississippi to the Invention of the Internet*. New York: The Free Press, 2001. 322 pp. (hc. 0-7432-1064-6) $40.00.
Gr. 10–12. This book is a tribute to the engineers whose work is evidenced in eight great projects. The book describes the engineers' vision, the projects, the controversies, and the challenges. The projects included taming the Mississippi River; building the Hoover Dam; in-

venting the light bulb; providing electricity to cities, towns, and farms; constructing the Croton Aqueduct; building Ammann's bridges; digging tunnels; and creating the Internet. At the end of the book there is an afterword, sources, and an index.

Chapter 7

Information Technology

Nonfiction books in this chapter contain information on the Internet and how it works and computers and how they work. There are books on email and netiquette. Some of the books provide a history of computing and introduce students to the men who contributed the logical concepts underlying computers including Alan Turing. Other book annotations in this chapter explore supercomputers, microchips, and virtual reality.

Eck, Michael. *The Internet: Inside and Out.* New York: PowerKids Press, 2002. 48 pp. (hc. 0-8239-6108-7) $18.95.
Gr. P–3. From the Technology: Blueprints of the Future series. This book explains in simple terms how the Internet works and describes the impact of the Internet on society. The book concludes with a section on the future, a glossary, additional resources, and an index.

Brimner, Larry Dane. *E-Mail.* New York: Children's Press, 1997. 47 pp. (hc. 0-516-20332-0) $22.00.
Gr. K–3. From a True Book series. Large font and colorful photographs entice young readers to learn about e-mail. The book explains how e-mail works, describes mailing lists, gives netiquette instructions, and has rules about the safe use of email. Resources for learning more, a glossary, and an index are included.

——. *The World Wide Web,* rev. ed. New York: Children's Press, 2000. 47 pp. (hc. 0-516-21935-9) $22.00.
Gr. K–3. From a True Book series. This basic book defines the World Wide Web, describes how it works, explains why it is special, and dis-

cusses ways children use the web. Resources for learning more, a glossary, and an index are included.

Drake, Jim. *What Is a Computer?* Chicago: Heinemann Library, 1999. 32 pp. (hc. 1-57572-787-0) $21.36.
Gr. 1–3. Photographs, large font, and simple text explain the inner workings of computers, peripherals, software, and storage devices. Words in bold font are defined in the glossary. This is a basic introduction to the computer for beginners. A glossary, more books to read, and an index are provided.

Lockman, Darcy. *The Internet.* Tarrytown, N.Y.: Marshall Cavendish, 2001. 48 pp. (hc. 0-7614-1046-5) $22.79.
Gr. 2–4. A brief description of the Internet is followed by an explanation of how the Internet is used for communication and research. A colorful diagram that explains hyperlinks is particularly useful. Full-page color photographs with captions add both appeal and additional information. A glossary, resources for finding out more, and an index conclude the book.

Jortberg, Charles A. *Virtual Reality and Beyond.* Edina, Minn.: Abdo and Daughters, 1997. 31 pp. (hc. 1-56239-728-1) $24.21.
Gr. 3–5. From the Kids and Computers series. This is a very brief survey of virtual reality. It includes a description of virtual reality, how it works, how it is used, and possible future uses. The book concludes with brief paragraphs and color photographs on careers in computer science. A glossary and an index conclude the book.

Jefferis, David. *Internet: Electronic Global Village.* New York: Crabtree, 2002. 32 pp. (hc. 0-7787-0052-6) $22.60; (pbk. 0-7787-0062-3) $17.25.
Gr. 4–8. The four sections of the book include a look at the early days of the Internet, an overview of the web, the global electronic village, and possibilities for the near future. Color photographs, diagrams, sidebars, and brief bursts of text provide a history of the Internet, describe innovations that made it possible, and explain the changes it has made in our lives. The reference section of the book contains a timeline, a glossary, and an index.

Baker, Christopher W. *Virtual Reality: Experiencing Illusion.* Brookfield, Conn.: The Millbrook Press, 2000. 48 pp. (hc. 0-7613-1350-8) $23.90.
Gr. 5–8. From the New Century Technology series. Enter the mysterious, fascinating world of virtual reality where things are not as they seem. This is a clear, succinct description of how virtual reality works and how it is presently being used. Sidebars with hands-on activities and color photographs help readers understand the concepts presented in the book. Resources for learning more and an index conclude the book.

Williams, Brian. *Computers.* Chicago: Heinemann Library, 2002. 48 pp. (hc. 1-58810-210-6) $25.64.
Gr. 5–8. From the Great Inventions series. From the abacus in 3000 B.C.E. to the eBook in 1999, clear concise text describes computer innovations. The descriptions are supplemented by a timeline, photographs, diagrams, and brief biographies of key people. A timeline of the history of computers, a glossary, more books to read, and an index are included.

Wolinsky, Art. *Communicating on the Internet.* Berkeley Heights, N.J.: Enslow, 1999. 64 pp. (hc. 0-7660-1260-3) $17.95.
Gr. 5–8. From the Internet Library series. There are different ways to communicate on the Internet including email, mailing lists, and chat rooms. This book provides a thorough examination of these means of communication. There is a chapter on netiquette that is particularly useful and it contains the ten commandments for Internet ethics. A glossary, books for further reading, and an index are provided.

——. *The History of the Internet and the World Wide Web.* Berkeley Heights, N.J.: Enslow, 1999. 64 pp. (hc. 0-7660-1261-1) $17.95.
Gr. 5–8. From the Internet Library series. This history of the Internet contains a great deal of technical information that may not be found in other books. Diagrams, photographs, and sidebars help explain the content and are useful additions. Throughout the text readers find a number of useful Internet sites. A glossary, books for further reading, and an index are provided.

Sherman, Josepha. *Jerry Yang and David Filo: Chief Yahoos of Yahoo!* Brookfield, Conn.: Twenty-First Books, 2001. 80 pp. (hc. 0-7613-1961-1) $23.90.

Gr. 5–12. From the Techies series. While working on doctorates in electrical engineering at Stanford University, Yang and Filo compiled a list of their favorite Internet sites and shared the list with their friends. Working on the list was more challenging than their studies and eventually they left the doctoral program to devote all of their time to their search engine, Yahoo. Black and white photographs, sources, and an index are included.

Billings, Charlene W. *Supercomputers: Shaping the Future.* New York: Facts On File, 1995. 132 pp. (hc. 0-8160-3096-0) $25.00.
Gr. 6–12. From the Facts On File Science Sourcebooks series. Supercomputers enable researchers to quickly analyze large amounts of data and they are used in research on the weather, the climate, the universe, and new materials. Throughout the book, readers are introduced to researchers whose inventions made computers and supercomputers possible. The book concludes with a chapter on supercomputers and virtual reality. A glossary, books for further reading, and an index conclude the book.

Grady, Sean M. *Virtual Reality: Computers Mimic the Physical World.* New York: Facts On File, 1998. 170 pp. (hc. 0-8160-3605-5) $25.00.
Gr. 6–12. From the Facts On File Science Sourcebooks series. The first chapters describe virtual realty, its capabilities, and the technology behind virtual reality. The next several chapters describe how virtual reality is used in science, in medicine, in the business world, in the classroom, and in the military. Information on drawbacks to virtual reality and the future of virtual reality conclude the book. Black and white photographs, a glossary, books for further reading, and an index are included.

Graham, Ian. *The Internet: The Impact on Our Lives.* Austin, Tex.: Raintree Steck-Vaughn, 2000. 64 pp. (hc. 0-7398-3173-9) $28.56.
Gr. 7–10. From the Twenty-First Century Debates series. Graham presents a balanced view of the pros and cons of the Internet. Freedom of information, e-commerce, and net crime are a few of the issues examined in the text. In the margins are text boxes containing facts, questions for debate, and quotes. These text boxes cause readers to pause and ponder the issues. A glossary, books to read, Internet addresses, and an index conclude the book.

——. *Internet Revolution*. Chicago: Heinemann Library, 2003. 64 pp. (hc. 1-40340-325-2) $28.50.
Gr. 7–10. From the Science at the Edge series. This history of the Internet describes the technology that makes it possible, the uses of the Internet, and the abuses of the Internet. Clear text, photographs, diagrams, and sidebars enhance the fact-filled book. A timeline, a glossary, books for further reading, and an index conclude the book.

Dunn, John M. *The Computer Revolution*. San Diego: Lucent, 2002. 112 pp. (hc. 1-56006-848-5) $27.45.
Gr. 7–12. From the World History series. From early calculating machines to future possibilities, this is a concise history of the computer and changes it has made in our lives. A timeline, photographs, and sidebars add impact to the text. Located at the end of the book are notes, sources for further reading, a bibliography, and an index.

McCormick, Anita Louise. *The Internet: Surfing the Issues*. Springfield, N.J.: Enslow, 1998. 128 pp. (hc. 0-89490-956-8) $20.95.
Gr. 7–12. From the Issues in Focus series. A brief history of the Internet begins this book that focuses on the possibilities and problems including who controls it and security issues. The book concludes with a look to the future. Material in this book can be used for thoughtful classroom discussions. Chapter notes, a glossary, resources for further reading, and an index are provided.

Ceruzzi, Paul E. *A History of Modern Computing*. Cambridge, Mass.: The MIT Press, 1998. 398 pp. (hc. 0-262-03255-4) $22.95.
Gr. 9–12. From the History of Computing series. Ceruzzi tells the story of computing from ENIAC in 1945 to the networking of office and homes in the 1990s. Throughout the book innovations, inventions, and patents mark the changes over time as computer technology developed. Notes, a bibliography, and an index conclude the book.

Davis, Martin. *The Universal Computer: The Road from Leibniz to Turing*. New York: W.W. Norton, 2000. 257 pp. (hc. 0-393-04785-7) $26.95.
Gr. 9–12. Martin describes computers as engines of logic and explains the logical concepts underlying computers. Along the way he introduces G. W. Leibniz, George Boole, Gottlob Frege, Georg Cantor, David Hilbert, Kurt Godel, and Alan Turing. Each of these men con-

tributed to the logical concepts that make computers work. An epilogue, notes, references, and an index conclude the book.

Reid, T. R. *The Chip: How Two Americans Invented the Microchip and Launched a Revolution.* New York: Random House Trade Paperbacks, 2001. 308 pp. (pbk. 0-375-75828-3) $13.95.
Gr. 9–12. The work of Jack Kilby and Robert Noyce's discovery led to the development of the silicon microchip for which Kibly received a Nobel Prize in Physics. Woven into the story is information about both inventors' extensive reading and how they problem-solve. An author's note, a note about sources, notes, and an index conclude the book.

Katz, Jon. *Geeks: How Two Lost Boys Rode the Internet Out of Idaho.* New York: Villard, 2000. 207 pp. (hc. 0-375-50298-X) $22.95.
Gr. 10–12. Jon Katz is a renowned author in the field of Internet technology and computers. He began this project researching two typical "computer nerds" or "geeks" as they are often called. He flew to Idaho to meet these young men and interview them. He became, unexpectedly, immersed in their lives and helped them see that they did have "choices" in what to do with their lives. This fascinating story is a winner because it is true and is told truthfully. The reader will be compelled to read the entire book to find out what happens to these two boys.

Winter, Paul A., and Mary E. Williams, eds. *Information Revolution: Opposing Viewpoints.* San Diego: Greenhaven, 1998. 202 pp. (hc. 1-56510-801-9) $32.45; (pbk. 1-56510-800-0) $16.85.
Gr. 10–12. This is an impressive collection of essays that offer differing viewpoints on compelling questions on topics including the benefits of the information revolution, the potential of the information revolution to transform education and work, and the threat to various rights posed by the revolution. The short, provocative essays can be the starting point for classroom discussions. A periodical bibliography, questions for further discussion, organizations to contact, a bibliography of books, and an index conclude the book.

Chapter 8

Media and Communications

Writing, the telegraph, the radio, the telephone, and the television are all forms of communication examined in the nonfiction books annotated in this chapter. These books contain information on the discoveries and inventions that led to different forms of communication as well as how they changed society. Some of the books contain descriptions and diagrams that show how the technology works.

Humphrey, Paul. *What Was It Like Before the Telephone?* Illus. by Lynda Stevens. Austin, Tex.: Raintree Steck-Vaughn, 1995. 32 pp. (pbk. 0-8114-3781-7) $5.45.
Gr. K–3. From the Read All about It series. This basic overview describes how messages were sent before the telephone was invented. In this book a group of children visit a science museum and discover early means of communication including the Pony Express, smoke signals, carrier pigeons, and the telegraph. A picture quiz and an index conclude the book.

Englart, Mindi Rose. *Music CDs From Start to Finish*. Illus. by Peter Casolino. Woodbridge, Conn.: Blackbirch Press, 2001. 32 pp. (hc. 1-56711-485-7) $24.95.
Gr. 2–4. Color photographs and concise text explain how a band makes a music CD. Thomas Edison and his phonograph are briefly mentioned. A glossary, resources for learning more, and an index conclude the book.

Birch, Beverly. *Marconi's Battle for Radio*. Illus. by Robin Bell Corfield. Hauppauge, N.Y.: Barron's Educational Series, 1995. 41 pp. (hc. 0-8120-6620-0) $10.00; (pbk. 0-8120-9792-0) $5.00.

Gr. 3–4. After building a radio station on the southwest coast of England, Marconi and his assistants traveled to the east coast of America. From this vantage point, they sent a wireless signal across the Atlantic to England. This book briefly tells the story of his struggles and ultimate success. Pastel illustrations aptly convey the challenges he encountered.

Merbreier, W. Carter, with Linda Capus Riley. *Television: What's Behind What You See*. New York: Farrar, Straus and Giroux, 1996. 40 pp. (hc. 0-374-37388-4) $16.00.

Gr. 3–5. Filled with cartoon-like illustrations and short paragraphs, the book explains the technology behind the television. Not only do the authors explain what happens in a television studio, they also explain how the signal travels from the studio to homes. The last two-page spread in the book has a television timeline and a look at the future of television.

Oxlade, Chris. *Telecommunications*. Austin, Tex.: Raintree Steck-Vaughn, 1997. 48 pp. (hc. 0-8172-4813-7) $27.12.

Gr. 3–6. From the Twentieth Century Inventions series. Photographs, clear succinct text, diagrams, and sidebars explain how telephones, telegraphs, radios, televisions, and computer networks operate. A timeline, a glossary, resources for learning more, and an index conclude the book.

Samoyault, Tiphaine. *Alphabetical Order: How the Alphabet Began*. New York: Viking, 1996. 32 pp. (hc. 0-670-87808-1) $14.99.

Gr. 3–6. Beginning with the premise that the alphabet is the most important innovation in human history, the book provides a clear, concise examination of the importance of alphabets throughout history. Samoyault introduces a variety of alphabets including symbol systems such as semaphores and Morse code. A glossary concludes the book.

Biel, Jackie. *Video*. Tarrytown, N.Y.: Marshall Cavendish, 1996. 63 pp. (hc. 0-7614-0048-6) $25.64.

Gr. 4–8. From the Inventors and Inventions series. The book describes the development of video technology and introduces the inventors and those who use the technology. One chapter explains how video works and then several chapters explain how video is used. A timeline, books for further reading, a glossary, and an index are included.

Holland, Gini, and Amy Stone. *Telephones*. Tarrytown, N.Y.: Marshall Cavendish, 1996. 63 pp. (hc. 0-7614-0065-6) $25.64.
Gr. 4–8. From the Inventors and Inventions series. This description of the history of the telephone explains that the only time phone companies have lost business was during the Depression. Not only does the book discuss innovations in the telephone, it also describes the telephone's impact on society. "Amazing Fact" boxes, photographs, reproductions, and brief biographies add to the text. A timeline, books for further reading, a glossary, and an index are included.

Riehecky, Janet. *Television*. Tarrytown, N.Y.: Marshall Cavendish, 1996. 63 pp. (hc. 0-7614-0045-1) $14.95.
Gr. 4–8. From the Inventors and Inventions series. Riehecky describes the development of television over the years and its impact on society. Diagrams explain how the technology works. "Amazing Fact" boxes, photographs, reproductions, and brief biographies add to the text. A timeline, books for further reading, a glossary, and an index are included.

Gearhart, Sarah. *The Telephone*. New York: Atheneum Books for Young Readers, 1999. 79 pp. (hc. 0-689-82815-2) $17.95.
Gr. 4–12. From the Turning Point Inventions series. From the past to the present to the future, readers discover the fascinating story of the telephone. Photographs and engaging text tell how events and circumstances in Alexander Graham Bell's life led to this invention that revolutionized communications. The book concludes with a list of books for further reading and an index.

Dowswell, Paul. *Entertainment*. Chicago: Heinemann Library, 2001. 48 pp. (hc. 1-58810-211-4) $25.64.
Gr. 5–8. From the Great Inventions series. This book is a chronological presentation of innovations in entertainment that is sure to capture readers' interest. Each innovation covers a two-page spread with a succinct description of the innovation, a brief timeline, photographs, and sidebars of additional information. The book concludes with a timeline that summarizes the innovations, a glossary, more books to read, and an index.

Williams, Brian. *Communications*. Chicago: Heinemann Library, 2002.
    48 pp. (hc. 1-58810-245-3) $25.64.
Gr. 5–8. From the Great Inventions series. Communication innovations
are examined in two-page spreads that contain just enough information
to get researchers started on their quest to learn more. Timelines, photographs, diagrams, and brief biographies in sidebars enhance the text.
The book ends with a timeline, a glossary, more books to read, and an
index.

Warburton, Lois. *The Beginning of Writing*. San Diego: Lucent, 1990.
    126 pp. (hc. 1-56006-113-8) $27.45.
Gr. 6–9. From the Overview series. From cave paintings to Sequoyah's
Cherokee "alphabet," this book contains a brief history of the development of writing enhanced by black and white illustrations. At the end of
the book are an epilogue, a glossary, a chronology, suggestions for
further reading, works consulted, and an index.

Sreffens, Bradley. *The Printing Press: Ideas into Type*. San Diego:
    Lucent, 1990. 96 pp. (hc. 1-56006-205-3) $28.70.
Gr. 6–12. From the Encyclopedia of Discovery and Invention series.
From ancient Chinese printing to Gutenberg's printing press to computers and laser printers, Sreffens traces the development of printing
placing it in a historical and a cultural context. Diagrams help explain
the printing processes used during different time periods. Color reproductions of illuminated manuscripts and black and white photographs
highlight the text. A glossary, books for further reading, a bibliography,
and an index are included.

Schwartz, Evan I. *The Last Lone Inventor: A Tale of Genius, Deceit,
    and the Birth of Television*. New York: HarperCollins, 2002. 322
    pp. (hc. 0-06-621069-0) $24.95.
Gr. 9–12. Lively, engaging prose tells the story of the patent battles
between Philo Farnsworth, the inventor of television, and David Sarnoff, the father of television. While Farnsworth was a lone, struggling
inventor, Sarnoff was head of the broadcasting giant Radio Corporation
of America (RCA) and had the resources to take control of the development of television. Notes and index conclude the book.

Standage, Tom. *The Victorian Internet: The Remarkable Story of the
    Telegraph and the Nineteenth Century's On-line Pioneers*. New

York: Penguin Putnam, 1998. 228 pp. (pbk. 0-425-17169-8) $12.00.
Gr. 9–12. In this history of the Victorian Internet, Standage traces the development of the telegraph in both England and the United States. He draws unmistakable parallels between the telegraph and today's Internet. Once the telegraph lines were strung criminals and lovers quickly found ways to use the telegraph to their advantage. Sources, an index, and the Morse code conclude the book.

Stashower, Daniel. *The Boy Genius and the Mogul: The Untold Story of Television*. New York: Random House, 2002. 277 pp. (hc. 0-7679-0759-0) $24.95.
Gr. 9–12. The preface for the book describes some hilarious misconceptions surrounding television when it was first introduced. Then the author informs readers that while scientists and research laboratories were struggling to create television, a fourteen-year-old boy, Philo T. Farnsworth, had already figured it out. This is the tale of Farnsworth's years of conflict with David Sarnoff, head of the RCA, over the patent rights to television. Black and white illustrations, a bibliography, and an index are included in this book.

Green, John. *The New Age of Communications*. New York: Henry Holt, 1997. 144 pp. (pbk. 0-8050-4027-7) $22.50.
Gr. 10–12. Green examines the digital revolution with a description of the convergences that have taken place as new technologies arrive and a look into future possibilities that are on the horizon. Information on artificial intelligence and virtual reality are included. A timeline, resources for further reading, and an index are provided.

Chapter 9

Medicine and Health

These books cover past, present, and future medical discoveries and inventions. From ether to vaccines, the breakthroughs discussed in these books affected the health of millions of people throughout the world. There are books on biotechnology, vaccines, infectious diseases, the history of dentistry, the uses of electricity in medicine, the greatest discoveries in medicine, and the greatest feuds in medicine.

Davis, Lucile. *Medicine in the American West*. New York: Children's Press, 2001. 32 pp. (hc. 0-516-22004-7) $21.00.
Gr. 3–5. From the Cornerstones of Freedom series. Most doctors practicing medicine in the American West received little formal training, lacked medicines, and did not have the equipment to effectively treat patients. The accepted medical practices often did more harm than good, so that even those with formal training were ill prepared to provide appropriate medical care. The book includes a glossary, a brief timeline, and an index.

Darling, David. *The Health Revolution: Surgery and Medicine in the Twenty-First Century*. Parsippany, N.J.: Simon & Schuster, 1996. 64 pp. (hc. 0-87518-616-5) $18.95; (pbk. 0-382-29170-5) $11.00.
Gr. 4–8. From the Beyond Two Thousand series. Magnetic resonance imaging, microsurgery, and cochlear implants are a few of the techniques changing medicine in the twenty-first century. Color photographs and diagrams help explain the medical techniques described in the text. A glossary, books for further reading, and an index are included.

125

Ichord, Loretta Frances. *Toothworms and Spider Juice: An Illustrated History of Dentistry*. Brookfield, Conn.: The Millbrook Press, 2000. 96 pp. (hc. 0-7613-1465-2) $24.90.
Gr. 4–8. This is a concise history of dentistry from ancient times to the present, complete with a variety of fascinating illustrations sure to cause readers to pause and take note. Source notes, a glossary, a bibliography, and an index are included.

Jefferis, David. *Biotech: Frontiers of Medicine*. New York: Crabtree, 2002. 32 pp. (hc. 0-7787-0051-8) $22.60; (pbk. 0-7787-0061-5) $17.25.
Gr. 4–8. Biotechnology refers to the linking of technology and biology, which combines science and engineering to improve living things. Examples of this are artificial organs, replacement parts for the body, and new foods. Illustrations, sidebars, diagrams, and intriguing text in a collage format make this an appealing book sure to be a favorite for browsing and discussing. A reference section, a timeline, a glossary, and an index conclude the book.

Parker, Steve. *Medical Advances*. Austin, Tex.: Raintree Steck-Vaughn, 1998. 48 pp. (hc. 0-8172-4896-X) $27.12.
Gr. 4–8. From the Twentieth Century Inventions series. From the prevention of illness to the diagnosis to treatment, this book contains a succinct overview of medical advances, augmented by color photographs and diagrams. The book concludes with a section on medicine of the future, a timeline, a glossary, books to read, and an index.

——. *Medicine*. New York: Dorling Kindersley, 1995. 64 pp. (hc. 0-56458-882-3) $15.99.
Gr. 4–8. From the Eyewitness Science series. This photographic history of medicine begins with a definition of medicine and ends with a look into the future. The book includes information on fads, fashions, and alternative treatments, which are not included in some books on the history of medicine. An index is included.

Donnellan, William L. *The Miracle of Immunity*. Tarrytown, N.Y.: Benchmark Books, 2003. 79 pp. (hc. 0-7614-1425-8) $28.50.
Gr. 5–8. From the Story of Science series. From ancient times to the present and into the future, this book contains useful research information for students studying the immune system. Photographs and dia-

grams are included. The book concludes with a glossary, books for further reading, and an index.

Dowswell, Paul. *Medicine*. Chicago: Heinemann Library, 2002. 48 pp. (hc. 1-58810-213-0) $25.64.
Gr. 5–8. From the Great Inventions series. Spanning the years from 3000 B.C. to 1982, this book contains information on a variety of medical innovations that had profound impacts on the practice of medicine. A two-page spread is devoted to each innovation and includes a timeline, photographs, diagrams, and a clear, concise description. A timeline, a glossary, more books to read, and an index are included.

Himrich, Brenda L., and Stew Thornley. *Electrifying Medicine: How Electricity Sparked a Medical Revolution*. Minneapolis: Lerner, 1995. 88 pp. (hc. 0-8225-1571-7) $23.93.
Gr. 5–8. Electricity keeps hearts beating, stimulates paralyzed muscles, eases pain, and stimulates hearing in the deaf. Color photographs and diagrams help explain the concepts discussed in the book. Information on new pathways for using electricity in medicine, a glossary, and an index conclude the book.

Miller, Brandon Marie. *Just What the Doctor Ordered: The History of American Medicine*. Minneapolis: Lerner, 1997. 88 pp. (hc. 0-8225-1737-X) $25.26.
Gr. 5–8. From the medicine of the Native Americans to the development of the polio vaccine, this is a very brief overview of medicine in America. The book contains quotes from primary sources, black and white illustrations, a selected bibliography, and an index.

Marrin, Albert. *Dr. Jenner and the Speckled Monster: The Search for the Smallpox Vaccine*. New York: Dutton Children's Books, 2002. 129 pp. (hc. 0-525-46822-2) $17.99.
Gr. 5–9. The book begins by describing the horrible ravages of smallpox and its history of destruction. Then, readers learn of Edward Jenner and his life's work to eradicate smallpox. In 1980, the World Health Organization (WHO) declared smallpox eliminated. All that remains are frozen vials of the variola to use for creating vaccine if terrorists use smallpox as a means of germ warfare. The book contains black and white photographs, a bibliography, Internet sources, and an index.

Altman, Linda Jacobs. *Plague and Pestilence: A History of Infectious
    Disease*. Berkeley Heights, N.J.: Enslow, 2001. 128 pp. (hc. 0-
    7660-1587-4) $20.95.
Gr. 6–9. From the Issues in Focus series. This brief history of infectious diseases also introduces the scientists who have discovered cures
and treatments for the diseases. The book ends with a discussion of new
diseases and yet to be discovered diseases that epidemiologists must be
prepared to prevent, cure, and control. Chapter notes, resources for
further reading, and an index are provided.

Woods, Michael, and Mary B. Woods. *Ancient Medicine: From Sor-
    cery to Surgery*. Minneapolis: Lerner, 2000. 86 pp. (hc. 0-8225-
    2992-0) $25.26.
Gr. 6–9. From the Ancient Technology series. Ancient medical treatments from the Stone Age, Egypt, India, China, Greece, and Rome are
described in this book. Photographs, a glossary, a bibliography, and an
index are included.

Yount, Lisa. *History of Medicine*. San Diego: Lucent, 2002. 128 pp.
    (hc. 1-56006-8-5-1) $27.45.
Gr. 6–9. From the World History series. This is a well-written, concise
history of medicine from the Stone Age to the present. Sidebars filled
with extensive quotes provide additional details that enhance the text. A
timeline, photographs, notes, a glossary, books for further reading, a
bibliography, and an index are included.

Farrell, Jeanette. *Invisible Enemies: Stories of Infectious Disease*. New
    York: Farrar, Straus and Giroux, 1998. 246 pp. (hc. 0-374-33637-
    7) $17.00.
Gr. 6–12. These stories are placed in the context of society and science
during the periods when these infectious diseases flourished. The diseases include smallpox, leprosy, the plague, tuberculosis, malaria, chol-
era, and AIDS. Photographs, drawings, and tidbits of information enhance the well-written, engrossing text. A glossary, books for further
reading, a selected bibliography, and an index are included in the book.

Hyde, Margaret O., and John F. Setaro. *Medicine's Brave New World*.
    Brookfield, Conn.: The Millbrook Press, 2001. 143 pp. (hc. 0-
    7613-1706-6) $29.90.
Gr. 6–12. With medial advances come moral and ethical dilemmas as
evidenced by the scenarios and case studies in this book. The authors

explore the issues of high-tech babies, spinal cord repair, creating spare human parts, stem cells, human cloning, and gene research. Clear, balanced explanations and thought-provoking questions provide material for lively classroom debates. A glossary, notes, resources for more information, and an index are provided.

Masoff, Joy. *Emergency*. Illus. by Brian Michaud and Peter Escobedo. New York: Scholastic, 1999. 48 pp. (hc. 0-590-97898-5) $16.95.
Gr. 6–12. This is a behind the scenes look at emergency medical care which explains the jobs of the people and how they use technology to save lives. Masoff explains how emergency room staff use common household objects such as plastic wrap, Krazy Glue, and a turkey baster. The book includes information on the evolution of emergency medical care, has brief notes on milestones in medicine, and a section on medical treatments of the future. Resources for learning more and an index are included.

Brynie, Faith Hickman. *Genetics and Human Health: A Journey Within*. Illus. by Sharon Lane Holm. Brookfield, Conn.: The Millbrook Press, 1995. 128 pp. (hc. 1-56294-545-9) $20.90.
Gr. 7–10. From Gregor Mendel's genetic research to present-day genetic research, this brief overview explains the basics of inherited diseases. Stories of real people impacted by inherited diseases hold the readers' attention. Charts, diagrams, and photographs aid in comprehending the text. Notes, a glossary, resources for further information, and an index are provided.

Nardo, Don. *Vaccines*. San Diego: Lucent, 2002. 127 pp. (hc. 1-56006-932-5) $27.45.
Gr. 7–10. From the Great Medical Discoveries series. Meet the medical researchers from Jenner to Pasteur to Salk to Sabin whose discoveries provided the world with vaccines. The book includes information on the search for an acquired immunodeficiency syndrome (AIDS) vaccine and cancer vaccines. Notes, a glossary, resources for further learning, and an index are provided.

Ritchie, David, and Fred Israel. *Health and Medicine*. New York: Chelsea House, 1995. 103 pp. (hc. 0-7910-2839-9) $21.95.
Gr. 7–10. From the Life One Hundred Years Ago Series. The book provides a concise examination of health care in the nineteenth century. Detailed captions accompany the photographs and reproductions,

which provide rich sources of additional information. Books for further reading and an index are included.

Cartright, Frederick F. *Disease and History,* 2nd ed. Illus. by Michael Biddiss. New York: Sutton Publishing, 2000. 220 pp. (hc. 0-7509-2315-6) $29.95.
Gr. 9–12. This book examines the impact of disease on society throughout the ages. While progress continues on physical and mental disorders, new ones continue to develop and to weaken nations. Some of these new disorders are created by modern technologies and hence the book includes information on possible future developments.

Fenster, Julie M. *Ether Day: The Strange Tale of America's Greatest Medical Discovery and the Haunted Men Who Made It.* New York: HarperCollins, 2002. 278 pp. (pbk. 0-06-019523-1) $12.95.
Gr. 9–12. This is the dark tale of three men, Horace Wells, William T. G. Morton, and Charles Jackson, who all claimed credit for inventing anesthesia and who were all destroyed by the ensuing battle. The graphic descriptions of the horror of surgery without anesthesia make readers aware of the importance of this medical discovery.

Friedman, Meyer, and Gerald W. Friedland. *Medicine's Ten Greatest Discoveries.* New Haven, Conn.: Yale University Press, 1998. 263 pp. (hc. 0-300-07598-7) $39.00.
Gr. 9–12. The ten most significant medical discoveries since 1543 include modern human anatomy, blood circulation, bacteria, vaccination, surgical anesthesia, x-ray beam, tissue culture, cholesterol, antibiotics, and DNA. The descriptions of these discoveries include information about the scientists mainly responsible for them. Notes and an index conclude the book.

Hellman, Hal. *Great Feuds in Medicine: Ten of the Liveliest Disputes Ever.* New York: John Wiley and Sons, 2001. 237 pp. (hc. 0-471-34757-4) $24.95.
Gr. 9–12. Scientists and their medical discoveries are subject to attacks, jealousy, and controversy. Hellman examines ten great medical feuds from blood circulation to AIDS. The book's content lends itself to lively discussions and spirited classroom debates. An epilogue, notes, a bibliography, and an index are included.

Ridley, Matt. *Genome: The Autobiography of a Species in Twenty-Three Chapters*. New York: HarperCollins, 2000. 320 pp. (pbk. 0-06-093290-2) $14.00.

Gr. 9–12. Ridley presents some of the results of the Human Genome Project by picking one gene from each pair of chromosomes and telling its story. This thought-provoking volume causes readers to reflect on the benefits of the research and the possible misuses of the research. A bibliography, notes, and an index are included.

Smolan, Rick, and Phillip Moffitt, eds. *Medicine's Great Journey: One Hundred Years of Healing*. Boston: Little, Brown, 1992. 172 pp. (hc. 0-8212-1987-1) $50.00.

Gr. 9–12. Dramatic, riveting photographs and concise, crisp text depict the history of medicine. Photographs of scientists and their breakthrough discoveries, doctors and nurses treating patients, and medical treatments show the very human side of medicine.

McCormick, Joseph B., and Susan Fisher-Hoch, with Leslie Alan Horvitz. *Level Four: Virus Hunters of the CDC*. New York: Barnes and Noble Books, 1999. 379 pp. (pbk. 0-7607-1211-5) $7.95.

Gr. 11–12. This husband and wife team wrote about their experiences with Level Four viruses in patients, in the field, and in the lab. Level Four viruses are lethal in humans and are ones for which there are limited treatments and no prevention. By writing the book, they hope to provide a greater understanding and awareness of these viruses and their impact on humans. An afterword, appendixes, and an index are provided.

Chapter 10

Science

This chapter includes books on the tools scientists use, simple machines, and the scientific revolution. Forensic science, the atom, DNA fingerprinting, genetic modification, and cloning are some of the topics encompassed in this chapter.

Fowler, Allan. *Simple Machines*. New York: Children's Press, 2001. 32 pp. (hc. 1-516-21680-5) $19.00; (pbk. 0-516-27310-8) $4.95.
Gr. P–2. From the Rookie Read-About Science series. Labeled photographs and simple sentences introduce young readers to simple machines. A glossary and an index conclude the book.

Walker, Sally M., and Roseann Feldman. *Inclined Planes and Wedges*. Illus. by Andy King. Minneapolis: Lerner, 2002. 48 pp. (hc. 0-8225-2221-7) $23.93.
Gr. 2–4. From the Early Bird Physics Books series. Readers learn about work, gravity, and friction and how they relate to inclined planes and wedges. Color photographs of young children using inclined planes and wedges help explain these simple machines to readers. A note to adults, resources for learning more, a glossary, and an index are included.

——. *Levers*. Illus. by Andy King. Minneapolis: Lerner, 2002. 48 pp. (hc. 0-8225-2220-9) $23.93; (pbk. 0-8225-2214-4) $7.95.
Gr. 2–4. From the Early Bird Physics Books series. Color photographs of young children exploring levers and easy to read text encourage children to learn about these simple machines. Hands-on activities using easy-to-find materials are depicted in the photographs and accom-

panying text. A note to adults, resources for learning more, a glossary, and an index conclude the book.

——. *Pulleys*. Illus. by Andy King. Minneapolis: Lerner, 2002. 48 pp. (hc. 0-8225-2220-9) $23.93; (pbk. 0-8225-2214-4) $7.95.
Gr. 2–4. From the Early Bird Physics Books series. Color photographs show pulleys at work and feature young children doing experiments to help them understand how pulleys work. A note to adults, resources for learning more, a glossary, and an index conclude the book.

——. *Screws*. Illus. by Andy King. Minneapolis: Lerner, 2002. 48 pp. (hc. 0-8225-2222-5) $23.93.
Gr. 2–4. From the Early Bird Physics Books series. Hands-on activities encourage students to explore with different screw configurations. The color photographs of young children and clear, simple directions encourage students to try the activities. A note to adults, resources for learning more, a glossary, and an index conclude the book.

——. *Wheels and Axles*. Illus. by Andy King. Minneapolis: Lerner, 2002. 48 pp. (hc. 0-8225-2219-5) $23.93.
Gr. 2–4. From the Early Bird Physics Books series. This book explains the parts of a wheel, an axle, and gears. Young readers are encouraged to experiment with wheels and axles. Color photographs and simple instructions show readers how to use simple materials to do the experiments. A note to adults, resources for learning more, a glossary, and an index conclude the book.

Giblin, James Cross. *The Mystery of the Mammoth Bones and How It Was Solved*. New York: HarperCollins, 1999. 97 pp. (hc. 0-06-027493-X) $15.95.
Gr. 4–6. In 1801 when word of the discovery of large old bones on a farm in New York State reached Charles Willson Peale, he immediately recognized that they would be an excellent addition to his natural history museum. The bones were a mystery waiting to be solved and when it was solved, the discovery caused scientists to rethink their theories of the age of the earth and eventually the theory of evolution. Located at the back of the book are a bibliography, source notes, and an index.

Markle, Sandra. *Science to the Rescue*. New York: Atheneum, 1994. 48 pp. (hc. 0-689-31783-2) $15.95.

Gr. 4–6. An introduction to the scientific method begins this book, which then gives readers a chance to apply the method by finding their own solutions to real life problems. Real life problems are presented and then technology devised to solve the problem is described. Next, readers are challenged to devise their own solutions to the problems. The hands-on experiments in the book require students to use the scientific method and think critically. The book concludes with an index.

Ross, Michael Elsohn. *Toy Lab*. Illus. by Tim Seely. Minneapolis: Carolrhoda Books, 2003. 48 pp. (hc. 0-87614-456-3) $23.93.
Gr. 4–6. By combining playing with favorite toys and the scientific method, Ross introduces youngsters to experiments in flight, pressure, gravity, and other scientific concepts. Interspersed between the experiments and questions are brief histories of the toys. A glossary and an index are included.

Richardson, Hazel. *How to Clone a Sheep*. Illus. by Andy Cooke. Danbury, Conn.: Franklin Watts, 2001. 92 pp. (hc. 0-531-14645-6) $14.50; (pbk. 0-531-16200-1) $4.95.
Gr. 4–8. This introduction to genetic research begins by describing clones and DNA. Information on scientists involved in genetic research and the ethical dilemmas caused by cloning are interesting reading and likely to encourage students to look for other resources in order to learn more. The activities, cartoon illustrations, and diagrams are fun and informative.

Levine, Shar, and Leslie Johnstone. *The Microscope Book*. Illus. by David Sovka. New York: Sterling, 1996. 80 pp. (hc. 0-8069-4898-1) $19.95; (pbk. 0-8069-4899-X) $10.95.
Gr. 5–8. This book introduces students to the care and use of a microscope. Simple experiments in biology, geology, forensic science, food, and the environment provide opportunities to practice using a microscope. Located at the end of the book are a glossary and an index.

Williams, Brian. *Science*. Chicago: Heinemann Library, 2002. 48 pp. (hc. 1-58810-242-9) $25.64.
Gr. 5–8. From the Great Inventions series. This book contains a look at carefully selected science innovations of interest to young people. The innovations are described across two-page spreads that feature photographs, diagrams, a brief timeline, and brief biographies of key indi-

viduals. A timeline, a glossary, a listing of more books to read, and an index conclude the book.

DuPrau, Jeanne. *Cloning*. San Diego: Lucent, 2000. 112 pp. (hc. 1-56006-583-4) $27.45.
Gr. 6–9. From the Overview series. The chapters contain discussions on cloning in agriculture, medicine, endangered species, and humans. Then, DuPrau discusses the moral and ethical issues regarding cloning. Black and white photographs and humorous cartoons add impact to the text. Notes, a glossary, resources for learning more, a bibliography, and an index are provided.

Goldstein, Natalie. *How Do We Know the Nature of the Atom*. New York: Rosen, 2001. 107 pp. (hc. 0-8239-3385-7) $26.50.
Gr. 6–9. From the Great Scientific Questions and the Scientists Who Answered Them series. Beginning with a set of basic questions about the atom, Goldstein tells stories about the scientists and their discoveries as they sought answers to the questions. Readers see how scientific discoveries evolved and how each scientist built on the work of others. A glossary, resources for learning more, and an index conclude the book.

Judson, Karen. *Genetic Engineering: Debating the Benefits and Concerns*. Berkeley Heights, N.J.: Enslow, 2001. 128 pp. (hc. 0-7660-1587-4) $20.95.
Gr. 6–9. From the Issues in Focus series. Judson carefully defines and describes genetic engineering. She then presents the benefits and concerns raised by genetic engineering in plants and animals. Genetic engineering and humans presents even more concerns, which are discussed, as are concerns for the future. Chapter notes, books for further reading, Internet addresses, a glossary, and an index are provided.

Henderson, Harry, and Lisa Yount. *The Scientific Revolution*. San Diego: Lucent, 1996. 111 pp. (hc. 1-56006-283-5) $27.45.
Gr. 6–12. From the World History Series. The authors trace the development of scientific discoveries between the years 1550 and 1900. Outer space, electricity, and the minute worlds discovered under microscopes are a few of the discoveries described in the book. Extensive quotes, photographs, and illustrations appeal to readers and help them focus on the material in the text. A timeline, notes, a glossary, books for further reading, a bibliography, and an index.

——. *Twentieth Century Science*. San Diego: Lucent, 1997. 112 pp. (hc. 1-56006-304-1) $27.45.
Gr. 6–12. From the World History Series. The book contains a brief overview of the inventions and discoveries of the twentieth century. including atoms, DNA, the solar system, artificial organs, and telecommunications. A timeline, black and white photographs, quotes, notes, a glossary, books for further reading, a bibliography, and an index are included.

Morgan, Sally. *Superfoods: Genetic Modifications of Foods*. Chicago: Heinemann Library, 2002. 64 pp. (hc. 1-58810-702-7) $19.50; (pbk. 1-4034-4123-5) $8.95.
Gr. 6–12. From the Science at the Edge series. Morgan introduces genetic modifications by describing a new type of rice that contains vitamin A, which has the potential to prevent blindness in six hundred thousand children a year. From crops to livestock, genetic modifications have been changing the food we eat, but these changes come with concerns. Morgan presents both sides of the issues and lets readers make their own decisions concerning the safety of genetic modifications. A timeline, a glossary, books for further reading, and an index are provided.

Spangenburg, Ray, and Diane K. Moser. *The History of Science From 1895 to 1945*. New York: Facts On File, 1994. 164 pp. (hc. 0-8160-2742-0) $25.00.
Gr. 6–12. From On the Shoulders of Giants series. An overview of life from the late 1880s to the mid-1900s is followed by a section on advances in the physical sciences and a section on advances in the life science. Exploring atoms, examining the universe, and studying genes are a just a few of the science discoveries discussed in this overview. An epilogue, an appendix on the scientific methods, a chronology, a glossary, books for further reading, and an index conclude the book.

——. *The History of Science From 1946 to 1990s*. New York: Facts On File, 1994. 176 pp. (hc. 0-8160-2743-9) $25.00.
Gr. 6–12. From On the Shoulders of Giants series. The first section discusses advances in the physical sciences including particles, quarks, and the solar system. The second section examines discoveries in the life science including DNA, RNA, and designer genes. An epilogue, a

chronology, a glossary, books for further reading, a list of resources, and an index conclude the book.

——. *Science and Invention*. New York: Facts On File, 1997. 158 pp. (hc. 0-8160-3402-9) $25.00.
Gr. 6–12. From the American Historic Places series. This carefully selected collection of historic building focus on those linked to science and invention in America. Each chapter focuses on a different site and begins with a brief introduction to the site and contact information. This is followed by a description of the site and why it is important. At the conclusion to the chapter are resources for learning more about the site. More places to visit, more reading sources, and an index are found at the end of the book.

Nardo, Don. *Cloning*. San Diego: Lucent, 2002. 127 pp. (hc. 1-56006-927-9) $27.45.
Gr. 7–10. From the Great Medical Discoveries series. A provocative introduction immediately discusses the controversy surrounding cloning. From the cloning of animals to the cloning of humans, Nardo presents the possible benefits. The book ends with a discussion of the moral and ethical issues surrounding cloning. Notes, resources for further reading, a bibliography, and an index are provided.

Suplee, Curt. *Milestones of Sci*ence. Washington, D.C.: National Geographic Society, 2000. 288 pp. (hc. 0-7922-7906-9) $35.00.
Gr. 7–12. This book contains an overview of scientific milestones that are sure to get students talking and involved in the book. The first five chapters include the classical era, the Middle Ages, the scientific revolution, the age of Newton, and the age of reason. The last four chapters examine physical sciences in the 1800s, life science in the 1800s, physical sciences in the 1900s, and life science in the 1900s. This oversized book contains illustrations, reproductions, and clear, color photographs. An index is included.

Brody, David Eliot, and Arnold R. Brody. *The Science Class You Wish You Had: The Seven Greatest Scientific Discoveries and the People Who Made Them*. New York: Penguin Putnam, 1997. 378 pp. (hc. 0-399-52313-98) $14.00
Gr. 9–12. The authors contend that survival depends on understanding science and with that end in mind have written about seven scientific discoveries they think everyone should understand. The discoveries

include gravity and physics; the atom; relativity; the Big Bang; evolution; cell and genetics; and DNA. Information is also included about the scientists who made the discoveries. Black and white illustrations accompany the very readable text. A chronology, a bibliography organized by chapters, and an index are provided.

Fridell, Ron. *DNA Fingerprints: The Ultimate Identity.* New York: Scholastic, 2001. 112 pp. (hc. 0-531-11858-4) $25.00.
Gr. 9–12. Readers learn about how DNA fingerprinting was discovered and how it is used in forensic identification. The book includes resources for learning more and an index.

——. *Solving Crimes: Pioneers of Forensic Science.* New York: Franklin Watts, 2000. 144 pp. (hc. 0-531-11721-9) $20.00.
Gr. 9–12. The pioneers profiled in this book include Alphonse Betrillon, Edward Henry, Karl Landsteiner, Edmond Locard, Clyde Snow, and Alec Jeffreys. These men persevered in their work even when faced with ridicule for the work they were doing. The book includes a glossary, black and white photographs, reproductions, diagrams, and resources for learning more.

Ne'eman, Yuval, and Yoram Kirsh. *The Particle Hunters,* 2nd ed. Cambridge, Mass.: Cambridge University Press, 1996. 298 pp. (hc. 0-521-47107-9) $85.00; (pbk. 0-521-47686-0) $30.00.
Gr. 10–12. Follow along as the authors uncover the atom's structure. This highly readable, informative text helps readers understand the basic concepts of particle physics and their impact on our lives. The book concludes with information on future trends, appendixes, a name index, and a subject index.

Suplee, Curt. *Physics in the Twentieth Century.* New York: Harry N. Abrams, 1999. 223 pp. (hc. 0-8109-4364-6) $49.50.
Gr. 10–12. Spectacular photographs and crisp text combine to tell the progress of physics in the twentieth century. Suplee explores the atom, the spectrum of "electromagnetic" radiation, quantum mechanics, structure of substances, the nucleus, chaos and order, and the cosmos. An index concludes the book.

Keynes, Richard Darwin. *Fossils, Finches, and Fuegians: Darwin's Adventures and Discoveries on the Beagle.* New York: Oxford University Press, 2003. 428 pp. (hc. 0-19-516649-2) $35.00.

Gr. 11–12. Written by Darwin's great-grandson, this book retells the famous voyage on the *H.M.S. Beagle* and traces the development of Darwin's thoughts on evolution. Maps, color plates of illustrations by artists on the *H.M.S. Beagle*, an epilogue, notes, and an index are included.

Chapter 11

Space

Discoveries and inventions that have made space travel possible and that have enabled scientists to probe the far reaches of the universe are described in the books annotated in this chapter. These books contain information on satellites, black holes, space ships, telescopes, and travel to the far reaches of space.

Branley, Franklyn M. *Floating in Space*. Illus. by True Kelly. New York: HarperCollins, 1998. 32 pp. (hc. 0-06-025432-7) $16.95; (pbk. 0-06-445142-9) $4.98.
Gr. K–3. From the Let's-Read-and-Find-Out-Science series. Cartoon-like drawings enhance the informative text that explains weightlessness in space and describes what it is like to live and work in space.

———. *The International Space Station*. Illus. by True Kelly. New York: HarperCollins, 1998. 32 pp. (hc. 0-06-028702-0) $16.95; (pbk. 0-06-445209-3) $4.98.
Gr. K–3. From the Let's-Read-and-Find-Out-Science series. Discover how the International Space Station (ISS) is being constructed in outer space. Learn about the research being conducted on the space station and what it is like to live on the ISS.

Graham, Ian. *The Best Book of Spaceships*. New York: Kingfisher, 1998. 33 pp. (hc. 0-7534-5133-6) $12.95.
Gr. K–3. Vivid, colorful illustrations that spill across two-page spreads and abbreviated text explain how rockets, probes, telescopes, and space shuttles work. A glossary and an index are included.

Gallant, Roy. *Space Stations*. Tarrytown, N.Y.: Marshall Cavendish,
    2000. 48 pp (hc. 0-7614-1035-X) $22.79.
Gr. 3–5. From the Kaleidoscope series. Gallant describes the Salyuts,
Skylab, Mir, and the International Space Station (ISS) in this brief history. Full-page color photographs and illustrations depict the space stations and the research and experiments being conducted in space. A
glossary, resources for learning more, and an index are included.

Walker, Niki. *Satellites and Space Probes*. New York: Crabtree, 1998.
    32 pp. (hc. 0-86505-681-1) $21.28; (pbk. 0-86505-691-9) $5.95.
Gr. 3–5. From the Eye on the Universe series. This basic introduction
to satellites and space probes tells what they are, how they are
launched, and how they are used. Colorful illustrations and diagrams
help explain key points discussed in the text. A glossary and an index
conclude the book.

Bergin, Mark. *Space Shuttle*. New York: Scholastic, 1999. 32 pp. (hc.
    0-531-14573-5) $27.50; (pbk. 0-531-15423-8) $9.95.
Gr. 3–6. Bright, colorful illustrations and abbreviated text describe the
building of the shuttle, lift off, living in space, and re-entry. Split pages
add to the fun of reading the book and provide different views of the
shuttle. The book includes a glossary, a list of shuttle facts, a chronology, and an index.

Farndon, John. *Rockets and Other Spacecraft*. Brookfield, Conn.: Cop-
    per Beech Books, 2000. 32 pp. (hc. 0-7613-1164-5) $21.90; (pbk.
    0-7613-0840-7) $14.95.
Gr. 3–6. From the How Science Works series. This brief survey explains how rockets and spacecraft are powered, how satellites orbit, and
what it is like to live in space. The pages are crowded with text, diagrams, cutaway views, and activities. A brief quiz and an index conclude the book.

Gifford, Clive. *How to Live on Mars*. Illus. by Scoular Anderson. New
    York: Franklin Watts, 2000. 96 pp. (hc. 0-531-14647-2) $14.50;
    (pbk. 0-531-16201-X) $4.95.
Gr. 3–6. Lively content laden writing and humorous black and white
line drawings describe what scientists and astronomers have learned
about Mars. The book looks to the future exploration and possible
colonization of Mars. One way to make it possible to live on Mars is by

terraforming, changing another planet to make it like earth. The book does not have a bibliography or an index.

Bergin, Mark. *Exploration of Mars*. New York: Franklin Watts, 2001.
    32 pp. (hc. 0-531-14615-4) $27.50; (pbk. 0-531-14807-6) $9.95.
Gr. 4–6. From the Fast Forward series. From the research of early astronomers to information obtained from Mars probes, the book begins with myths and facts about Mars. Then the book moves into the future and discusses what it will take to explore and live on Mars. A glossary, Mars facts, a chronology, and an index are provided.

Couper, Heather, and Nigel Henbest. *Black Holes*. New York: DK
    Publishing, 1996. 45 pp. (hc. 0-78940451-6) $10.95.
Gr. 4–6. Diagrams, illustrations, photographs, and brief bursts of text introduce readers to black holes and to the scientists whose theories explain them. While the format of the book is cluttered, the concepts intrigue students who are interested in the mysteries of our solar system. A glossary and an index conclude the book.

Spangenburg, Ray, and Kit Moser. *Artificial Satellites*. New York:
    Franklin Watts, 2001. 127 pp. (hc. 0-531-11760-X) $33.50; (pbk.
    0-531-13971-9) $11.65.
Gr. 4–8. From the Out of this World series. This overview of artificial satellites begins with the launch of Sputnik in 1957. Charts to help organize the information accompany the descriptions of the different types of satellites and their uses. A timeline, a glossary, resources to find out more, and an index conclude the book.

Bond, Peter. *Guide to Space*. New York: DK Publishing, 1999. 64 pp.
    (hc. 0-7894-3946-8) $19.95.
Gr. 5–8. Examine these spectacular color photographs beamed back to Earth from the Hubble Space Telescope and other robotic space probes. Readers learn what astronomers are discovering in these rich sources of data. Space data and an index are provided.

Dyson, Marianne J. *Space Station Science: Life in Free Fall*. New
    York: Scholastic, 1999. 128 pp. (hc. 0-590-05889-4) $16.95.
Gr. 5–8. Dyson gives an insider's look at astronaut training; living and working in space; and reentry. Details about past missions and quotes from the astronauts combine to make this a very readable text. Color photographs, diagrams, and simple activities help readers understand

what it is like aboard the space station. A glossary, an index, and resources for further study are included.

Fox, Mary Virginia. *Rockets*. Tarrytown, N.Y.: Marshall Cavendish, 1996. 63 pp. (hc. 0-7614-0063-X) $25.64.
Gr. 5–8. From the Inventors and Inventions series. From the earliest Chinese rockets to the present, this is a concise history of rockets. Diagrams, photographs, sidebars, and biographies add to the information presented in the text. A timeline, books for further reading, a glossary, and an index are provided.

——. *Satellites*. Tarrytown, N.Y.: Marshall Cavendish, 1996. 63 pp. (hc. 0-7614-0049-4) $25.64.
Gr. 5–8. From the Inventors and Inventions series. Fox describes the development of satellites and explains how they are used for things such as predicting weather and spying. Color photographs, diagrams, and biographies of scientists and astronauts add to this informative text. A timeline, books for further reading, a glossary, and an index are provided.

Williams, Brian. *Space*. Chicago: Heinemann Library, 2002. 48 pp. (hc. 1-58810-243-7) $25.64.
Gr. 5–8. From the Great Inventions series. Travel and research in space led to the innovations showcased in this book. A two-page spread with a description, photographs, diagrams, and brief biographies is allocated to each innovation. The book concludes with a timeline, a glossary, more books to read, and an index.

Spangenburg, Ray, and Kit Moser. *Onboard the Space Shuttle*. New York: Franklin Watts, 2002. 112 pp. (hc. 0-531-11896-7) $33.50; (pbk. 0-531-15568-4) $14.95.
Gr. 5–9. From the Out of This World series. Clearly written text and color photographs portray the history of the space shuttle. The authors describe the work done by the astronauts and describe the technology that makes living and working in space possible. A portion of the text is devoted to the *Challenger* tragedy. A chart of key shuttle flights, a space shuttle timeline, a glossary, resources for learning more, and an index are provided.

Vogt, Gregory L. *Spacewalks: The Ultimate Adventure in Orbit.* Berkeley Heights, N.J.: Enslow, 2000. 48 pp. (hc. 0-7660-1305-7) $18.95.

Gr. 5–9. From the Countdown to Space series. Space walkers spend a great deal of time preparing for their walks with underwater training, inside a simulator, and using virtual reality hardware. This well-written text filled with anecdotes and facts keeps readers intrigued from beginning to end. Photographs, chapter notes, a glossary, books for further reading, and an index are included.

MacDonald, Olive. *What If We Lived on Another Planet?* New York: Children's Press, 2002. 48 pp. (hc. 0-516-23912-0) $20.00; (pbk. 0-516-23479-X) $6.95.

Gr. 5–10. From the What If series. Explore the planets and then learn what it would take to live on another planet. MacDonald describes the current research and innovations that may one day make it possible to live on another planet. This high-interest book contains a map, photographs, and illustrations. A glossary, books for further reading, Internet sites, and an index are provided.

Stott, Carole. *Moon Landing: The Race for the Moon.* Illus. by Richard Bonson. New York: DK Publishing, 1999. 48 pp. (hc. 0-7894-3958-1) $14.95.

Gr. 5–10. Diagrams and photographs explore the technology that made travel to the moon possible and enhance this overview of space travel and moon landings. Included in the book is a futuristic illustration of a colony on the moon. Moon data, a list of moon missions, and an index conclude the book.

Wunsch, Susi Trautmann. *The Adventures of Sojourner: The Mission to Mars that Thrilled the World.* New York: Mikaya Press, 1998. 64 pp. (hc. 0-9650493-5-3) $18.40; (pbk. 0-9650493-6-1) $9.95.

Gr. 5–10. This is a fascinating, factual account of the Mars Pathfinder mission and the remote controlled Sojourner that traveled across Mars' surface sending back photographs and data about the soil, rocks, and climate. Photographs and drawings capture the excitement and adventure of this mission. A timeline of Mars voyages, a facts chart comparing Mars and Earth, an index, and Mars websites conclude the book.

Villard, Raymond. *Large Telescopes: Inside and Out.* Illus. by Alessandro Bartolozzi, Leonello Calvetti, and Lorenzo Cecchi. New

York: Power Plus Books, 2002. 48 pp. (hc. 0-8239-6110-9) $26.50.

Gr. 5–12. From the Technology: Blueprints of the Future series. The wonders of the heavens are found by peering through the lens of telescopes. Awe-inspiring color photographs, drawings, and diagrams fill the pages of this book that describes the first telescopes, the advances made in telescopes through the years, and telescopes of the future. A glossary, additional resources, and an index are included.

Cobb, Allan B. *How Do We Know How Stars Shine*. New York: Rosen, 2001. 107 pp. (hc. 0-8239-3380-6) $26.50.

Gr. 6–9. From the Great Scientific Questions and the Scientists Who Answered Them series. Beginning with the ancient Egyptians and Greeks' astronomy studies, Cobb traces the study of the stars through the ages. He introduces scientists and shows how their research builds on the research of other scientists. A glossary, resources for learning more, and an index conclude the book.

Fradin, Dennis Brindell. *The Planet Hunters: The Search for Other Worlds*. New York: Margaret K. McElderry Books, 1997. 148 pp. (hc. 0-689-81323-6) $19.95.

Gr. 7–12. Journey back in time and through space to explore the discoveries of the nine known planets, planets orbiting distant pulsars, and the possible tenth planet. A list of numbers used in the book, metric measurements, a chart on the nine known planets of the solar system, a bibliography, and an extensive biography are provided.

Crouch, Tom D. *Aiming for the Stars: The Dreamers and Doers of the Space Age*. Washington, D.C.: Smithsonian Institution Press, 1999. 338 pp. (hc. 1-56098-386-8) $29.95.

Gr. 9–12. In this narrative account of the space age, readers meet the dreamers and doers whose work and inventiveness made space travel possible. History, politics, and technology are woven together to describe the exploration of space. Black and white illustrations, notes, a selected bibliography, and an index are included.

McNab, David, and James Younger. *The Planets*. New Haven, Conn.: Yale University Press, 1999. 240 pp. (hc. 0-300-08044-1) $35.00.

Gr. 9–12. This book is a companion to the A and E television series by the same name. It introduces the planets, the people who discovered the planets, and describes what has been learned about the planets. Photographs, reproductions, and sidebars highlight information in the text. The book concludes with a chronology, statistics and facts about the planets, books for further reading, websites, and an index.

Heppenheimer, T. A. *The Space Shuttle Decision, 1965–1972. History of the Space Shuttle, Vol. 1.* Washington, D.C.: Smithsonian Institution Press, 2002. 470 pp. (pbk. 1-58834-014-7) $22.95.
Gr. 10–12. From the National Aeronautics and Space Administration (NASA) archives comes this examination of the decision to develop the space shuttle including the political maneuverings, the engineering issues, and the budget battles. A list of abbreviations and acronyms, black and white photographs and diagrams, a bibliography, and an index are found in the book.

——. *Development of the Space Shuttle, 1972–1981. History of the Space Shuttle, Vol. 2.* Washington, D.C.: Smithsonian Institution Press, 2002. 480 pp. (pbk. 1-58834-009-0) $22.95.
Gr. 10–12. The NASA archives provided the resources used in this description of the challenges and successes encountered in the development of the space shuttle. A list of abbreviations and acronyms, black and white photographs and diagrams, a bibliography, and an index are found in the book.

Parker, Barry. *Einstein's Brainchild: Relativity Made Relatively Easy!* Illus. by Lori Scoffield-Beer. Amherst, N.Y.: Prometheus Books, 2000. 280 pp. (hc. 1-57392-857-7) $28.00.
Gr. 10–12. Parker provides a simplified explanation of Einstein's theories enhanced by cartoons and diagrams. Information on space-time, time travel, gravity, and black holes are included in the book. An epilogue, notes, a glossary, a bibliography, and an index conclude the book.

Chapter 12

Transportation

Flying through the air, speeding along tracks, driving down the road, sailing over the ocean, or pedaling down the sidewalk involve different forms of transportation. The selected books annotated below explore the discoveries and inventions that make transportation possible.

Gibbons, Gail. *Bicycle Book.* New York: Holiday House, 1995. 32 pp. (hc. 0-8234-1199-0) $15.95.
Gr. P–2. Beginning with Leonardo da Vinci's sketch of a bicycle and the first bicycle called a "hobby horse," Gibbons presents a brief examination of the development of the bicycle. The book contains information on taking care of bikes and safety rules for riding bikes.

Baum, Brian. *Super Jumbo Jets: Inside and Out.* New York: PowerKids Press, 2002. 48 pp. (hc. 0-8239-6112-5) $18.95.
Gr. P–3. From the Technology: Blueprints of the Future series. Readers peek inside super jumbo jets and learn about the technology and scientific discoveries that enable them to fly. The book concludes with a section on the future, a glossary, additional resources, and an index.

Biello, David. *Bullet Trains: Inside and Out.* New York: PowerKids Press, 2002. 48 pp. (hc. 0-8239-6113-3) $18.95.
Gr. P–3. From the Technology: Blueprints of the Future series. Discover the technology that enables the bullet train to travel at half the speed of a jet plane. The book concludes with a section on the future, a glossary, additional resources, and an index.

Simon, Seymour. *Seymour Simon's Book of Trucks*. New York: HarperCollins, 2000. 32 pp. (hc. 0-06-028473-0) $18.99.
Gr. P–3. Full-page color photographs depict the big trucks young readers see on the highway such as a semitrailer truck, a dump truck, and a cement mixer. Just the right amount of text is used to explain what the truck does without overwhelming young readers with details.

Budd. E. S. *Fire Engines*. Eden Prairie, Minn.: The Child's World, 1998. 24 pp. (hc. 1-56766-656-6) $21.36.
Gr. K–3. Colorful, close-up photographs and short simple sentences in a large font describe fire trucks and explain how they work. A glossary contains the definitions of the words in bold print found throughout the text.

Richards, Jon. *Trains*. Illus. by Simon Tegg. Brookfield, Conn.: The Millbrook Press, 1998. 32 pp. (hc. 0-7613-0743-5) $17.95.
Gr. 1–3. Bright colorful photographs, cutaway diagrams, and abbreviated text tell the history of trains, explain how they work, describe different kinds of trains, and discuss how they are used. An index concludes the book.

Stille, Darlene. *Helicopters*. New York: Children's Press, 1997. 47 pp. (hc. 0-516-20335-5) $23.50; (pbk. 0-516-26171-1) $8.95.
Gr. 1–3. From a True Book series. This brief introduction to helicopters explains how they operate and how they are used. Photographs, resources for finding out more, a glossary, and an index are included.

Dolan, Edward F. *The Transcontinental Railroad*. Tarrytown, N.Y.: Marshall Cavendish, 2003. 32 pp. (hc. 0-7614-1455-X) $22.79.
Gr. 2–4. From the Kaleidoscope series. This is a very concise survey of the building of the transcontinental railroad complete with photographs and illustrations. Dolan uses simple language and descriptions to help readers understand the material. A glossary, a timeline, and an index conclude the book.

Krensky, Stephen. *Taking Flight: The Story of the Wright Brothers*. Illus. by Larry Day. New York: Simon & Schuster, 2000. 43 pp. (hc. 0-689-81225-6) $15.00.
Gr. 2–4. From the Read-to-Read Simon and Schuster Books for Young Readers series. This short chapter book chronicles the Wright brothers' years of work on their flying machine. Color illustrations fill the pages

and provide the support young readers need to enhance their comprehension of the story of flight.

Otfinoski, Steven. *Behind the Wheel: Cars Then and Now*. Tarrytown, N.Y.: Marshall Cavendish, 1997. 32 pp. (hc. 0-7614-0403-1) $17.95.
Gr. 2–4. Vivid color photographs of vintage cars from 1885 to the present and short descriptive sentences in large print provide a survey of the history of the automobile. Resources for learning more and an index conclude the book.

Busby, Peter. *First to Fly: How Wilbur and Orville Wright Invented the Airplane*. Illus. by David Craig. New York: Crown, 2002. 32 pp. (hc. 0-375-81287-3) $19.95.
Gr. 3–5. Photographs, drawings, reproductions, diagrams, and sidebars enhance this marvelous story of two dedicated men who pursued their childhood interest in flight throughout their lifetimes. A timeline, a glossary, a selected bibliography, and an index are included.

Weitzman, David. *Jenny: The Airplane that Taught America to Fly*. Brookfield, Conn.: Roaring Brook Press, 2002. 40 pp. (hc. 0-7613-1547-0) $17.95.
Gr. 3–5. Three children gather around their grandmother to hear how their great-grandmother helped to build and learned to fly the JN4–D airplane. The "Jenny" was the first mass-produced airplane and was used in World War II. Many pilots including Amelia Earhart and Charles Lindbergh learned to fly in these planes. Detailed black and white drawings show how the plane was constructed. The endpapers have a labeled cutaway view of the Jenny.

——. *Locomotive: Building An Eight-Wheeler*. Boston: Houghton Mifflin, 1999. 40 pp. (hc. 0-395-69687-9) $16.00.
Gr. 3–5. Detailed pen and ink drawings show a steam locomotive being built in 1870. Readers follow along from the drafting room to the machine shop to the forge and finally to the erecting shop. The informative text not only describes the process, but also provides information on the history of the railroad.

——. *Model T: How Henry Ford Built a Legend*. New York: Crown, 2002. 40 pp. (hc. 0-375-81107-9) $18.99.

Gr. 3–5. Weitzman chronicles Henry Ford's development of the Model T including the assembly line and other innovations that made the car affordable to the masses. Black and white drawings span two-page spreads and depict the workers as they assembled the cars. Interesting bits of information about the automobile are scattered in the captions of the illustrations.

Kirkwood, Jon. *The Fantastic Cutaway Book of Giant Machines*. Illus. by Alex Pang. Brookfield, Conn.: The Millbrook Press, 1996. 40 pp. (hc. 0-7613-0498-3) $20.95.
Gr. 3–6. Look inside to discover the inner workings and the technology required to operate big machines such as the iron horse, helicopters, trucks, submarines, and cars. Bright colorful photographs, illustrations, and cutaway diagrams fill the pages of the book accompanied by short paragraphs of text.

Blumberg, Rhoda. *Full Steam Ahead: The Race to Build a Transcontinental Railroad*. Washington, D.C.: National Geographic Society, 1996. 159 pp. (hc. 0-7922-2715-8) $18.95.
Gr. 4–6. Meet the scheming businessmen, the irate Native Americans, and the mistreated Chinese immigrant laborers who all played a part in the construction of the transcontinental railroad in this well-researched book. Black and white photographs and reproductions accompany the decisive text. Notes, a bibliography, and an index are included.

De Angelis, Gina. *The Hindenburg*. Philadelphia: Chelsea House, 2001. 119 pp. (hc. 0-7910-5272-9) $22.95.
Gr. 4–6. From the Great Disasters series. Quotes from primary sources and vintage photographs add to this history of the Hindenburg and other airships. The last chapter in the book examines possible future uses for airships. An airship chronology, books for further reading, and an index are included.

Wilkinson, Philip. *Ships*. New York: Kingfisher, 2000. 64 pp. (hc. 0-7534-5280-4) $16.95.
Gr. 4–6. This is an entertaining and informative guide to boats, ships, galleons, battleships, and other oceangoing vessels. Maps, detailed drawings, photographs, and cutaway diagrams cover the two-page spreads. Throughout the text, readers learn about discoveries and inventions that make life on the seas safer and enable researchers to explore the sea.

Harrison, Peter. *Cars*. New York: Anness Publishing, 2001. 64 pp. (hc. 0-7548-0628-6) $14.95.
Gr. 4–7. From the Investigations series. Color photographs, diagrams, and concise text explain the science involved in automobile mechanics and provide information about car design, car racing, and car models. A glossary and an index conclude the book.

Carson, Mary Kay. *The Wright Brothers: How They Invented the Airplane with Twenty-One Activities Exploring the Science and History of Flight*. Illus. by Laura D'Argo. Chicago: Chicago Review Press, 2003. 160 pp. (pbk. 1-55652-477-3) $14.95.
Gr. 4–8. More than a biography, this book provides students with hands-on activities that help them understand the forces of lift, thrust, gravity, and drag. An understanding of these four forces was necessary for the Wright brothers to succeed in flying their airplanes. The book includes photographs, diagrams, illustrations, and clear detailed instructions for the twenty-one activities.

Hansen, Ole Steen. *Amazing Flights: The Golden Age*. New York: Crabtree, 2003. 32 pp. (hc. 0-7787-1202-8) $17.94; (pbk. 0-7787-1218-4) $8.95.
Gr. 4–8. From the Story of Flight series. Across the Atlantic, across the Pacific, and across the North Pole are some of the daring exploits described in this book. Readers learn about Charles Lindbergh, James Doolittle, barnstormers, and daredevils' flights that enthralled the public during the golden age of aviation. Illustrations, a plane identification guide, a glossary, and an index are included.

——. *Commercial Aviation*. New York: Crabtree, 2003. 32 pp. (hc. 0-7787-1205-2) $17.94; (pbk. 0-7787-1221-4) $8.95.
Gr. 4–8. From the Story of Flight series. This book explains the development of planes and engines that made commercial aviation a viable enterprise. A glimpse into the future of airliners, a spotters' guide, illustrations, a glossary, and an index are included.

——. *The Wright Brothers and Other Pioneers of Flight*. New York: Crabtree, 2003. 32 pp. (hc. 0-7787-1200-1) $17.94.
Gr. 4–8. From the Story of Flight series. Meet a variety of pioneers who led the race to fly including among others Otto Lilienthal, Glenn Curtis, and Petr Nesterov. A plane identification guide, illustrations, a glossary, and an index are included.

Murdico, Suzanne J. *Concorde*. New York: Children's Press, 2001. 48
pp. (hc. 0-516-23158-8) $20.00; (pbk. 0-516-23261-4) $6.95.
Gr. 4–8. From the Built for Speed series. Learn about the designers and
engineers and the challenges they overcame as they created this super-
sonic transport. Photographs, diagrams, and a fact sheet on the Concorde are included. A glossary, books for further reading, resources for
learning more, and an index conclude the book.

Rinard, Judith E. *The Story of Flight: The Smithsonian National Air
and Space Museum*. New York: Firefly Books, 2002. 64 pp. (hc. 1-
55297-642-4) $16.95; (pbk. 1-55297-694-7) $8.95.
Gr. 4–8. Pictures with lively captions plus engaging, interesting text tell
the story of flight. This is a compact, succinct history of aviation based
on the National Air and Space Museum collection. An index concludes
the book.

Dowswell, Paul. *Transportation*. Chicago: Heinemann Library, 2001.
48 pp. (hc. 1-58810-216-5) $25.64.
Gr. 5–8. From the Great Inventions series. Two-page spreads featuring
innovations in transportation include a succinct description of the innovation, a timeline, photographs, and diagrams. The book concludes
with a timeline that summarizes the innovations, a glossary, more
books to read, and an index.

Fisher, Leonard Everett. *Tracks across America: The Story of the
American Railroad*. New York: Holiday House, 1992. 192 pp. (hc.
0-8234-0945-7) $19.95.
Gr. 5–12. This is an exciting history of the development of the railroad
in America during the nineteenth century. The immigrants who built
the railroad, the Native Americans who resisted the construction, and
the impact of the railroad during the Civil War are vividly described.
Photographs, drawings, a selected bibliography, and an index are included.

Barter, James. *Building the Transcontinental Railroad*. San Diego: Lucent, 2002. 127 pp. (hc. 1-56006-880-9) $27.45.
Gr. 6–8. From the World History series. A timeline of important dates
at the beginning of the book presents an overview of the material,
which aids students' comprehension. Barter's engaging, descriptive
narrative tells the colorful history of the transcontinental railroad.
Maps, sidebars, quotes, photographs, and illustrations aid in under-

standing the text. Notes, resources for further reading, a bibliography, and an index conclude the book.

Cefrey, Holly. *High Speed Trains*. New York: Children's Press, 2001. 48 pp. (hc. 0-516-23517-X) $17.25; (pbk. 0-516-23260-6) $6.95.
Gr. 6–8. From the Built for Speed series. While high-speed trains are not familiar to most Americans they are familiar to people in other countries and Cefrey describes the high-speed trains found in other countries. The book begins with a brief history of the development of trains and then explains the technology needed for high-speed trains operate. A glossary, resources for learning more, and an index are included.

Cotter, Allison. *Cycling*. San Diego: Lucent, 2002. 127 pp. (hc. 1-59018-071-2) $27.45.
Gr. 6–8. From the History of Sports series. This history of cycling includes a chapter on the invention of the bicycle and a chapter on how the bicycle impacted society. The evolution of the bicycle from a mode of transportation to a recreational activity for much of the world is discussed. Awards and statistics, notes, books for further reading, a bibliography, and an index conclude the book.

Woods, Michael, and Mary B. Woods. *Ancient Transportation: From Camels to Canals*. Minneapolis: Lerner, 2000. 95 pp. (hc. 0-8225-2993-9) $25.26.
Gr. 6–8. From the Ancient Technology series. Step back in time to discover how ancient civilizations developed different modes of transportation, built bridges, and drew maps. Photographs, a glossary, a bibliography, and an index are included.

Tunis, Edwin. *Wheels: A Pictorial History*. Baltimore, Md.: Johns Hopkins University Press, 2002. 96 pp. (hc. 0-8018-6926-3) $24.95.
Gr. 6–12. This visual history of wheels is also the history of transportation was originally published in 1955. Engaging narrative and detailed black and white drawings describe the evolution of transportation across time and across nations.

Crouch, Tom D., and Peter L. Jakab. *The Wright Brothers and the Invention of the Aerial Age*. Washington, D.C.: National Geographic Society, 2003. 240 pp. (hc. 0-7922-6985-3) $35.00.

Gr. 7–12. Over two hundred black and white photographs, quotes from the Wright brothers, and engaging narrative tell the story of the aerial age. A bibliography and an index conclude the book.

Rinard, Judith E. *The Book of Flight: The Smithsonian Institution's National Air and Space Museum.* New York: Firefly Books, 2001. 128 pp. (hc. 1-55209-619-X) $24.95; (pbk. 1-55209-599-1) $14.95.
Gr. 7–12. From the Wright brothers to the astronauts and from airships to flying boats, flight fascinates and intrigues us. Color photographs, reproductions, and diagrams tell the story of major milestones in air travel. Fun facts appear in small text boxes throughout the book. A timeline, a glossary, and an index conclude the book.

Schafer, Mike, with Joe Welsh and Kevin Holland. *The American Passenger Train.* St. Paul, Minn.: Motorbooks, 2001. 156 pp. (hc. 0-7603-0896-9) $34.95.
Gr. 7–12. Filled with color photographs and reproductions, this is a history of the American passenger train. From cars made of wood to cars made of steel and from cars lit by gas to cars lit by electricity, the passenger trains incorporated new technologies as they became available. Sidebars and an index are included.

Bilstein, Roger E. *The Enterprise of Flight: The American Aviation and Aerospace Industry.* Washington, D.C.: Smithsonian Institution Press, 2001. 280 pp. (pbk. 1-56098-964-5) $19.95.
Gr. 9–12. The development of passenger planes, small planes, business jets, helicopters, military aircraft, and the International Space Station from the 1900s to the present are all described in this compact history. The impact of international competition on the industry is examined. An appendix, a chronology, notes, a selected bibliography, and an index conclude the book.

Coffey, Frank, and Joseph Layden. *America on Wheels: The First One Hundred Years: 1896–1996.* Los Angeles: General Publishing Group, 1996. 304 pp. (hc. 1-881649-80-6) $40.00.
Gr. 9–12. This oversized book is filled with photographs that document America's passion for the automobile. The text presents a broad overview of the history of the automobile in America. This book is a companion to the Public Broadcasting Service (PBS) special. A bibliography and an index conclude the book.

Crouch, Tom D. *A Dream of Wings: America and the Airplane, 1875–1905.* New York: W.W. Norton, 2002. 349 pp. (pbk. 0-393-32227-0) $14.95.

Gr. 9–12. The Wright brothers' successful flight was the culmination of years of attempts at flight by engineers, scientists, and dreamers. This book tells their stories of success and failure. An epilogue, notes, a bibliography, and an index conclude the book.

Chapter 13

Weapons and Warfare

From ancient times to the present, the books annotated below chronicle the evolution of weapons and warfare. From hand weapons such as the ax and sword of ancient times to remote-controlled airborne weapons these books include a vast array of weapons. Advances in weapons and warfare used on the land, air, and sea are described in these books.

Beyer, Mark. *Aircraft Carriers: Inside and Out*. New York: PowerKids Press, 2002. 48 pp. (hc. 0-8239-6117-7) $18.95.
Gr. P–3. From the Technology: Blueprints of the Future series. Explore an aircraft carrier from its construction to its becoming a floating city. The book concludes with a section on the future, a glossary, additional resources, and an index.

Budd, E. S. *Tanks*. Chanhassen, Minn.: The Child's World, 2002. 24 pp. (hc. 1-56766-984-0) $21.36.
Gr. K–2. Large color photographs, large font, and short simple sentences explain what tanks are and how they work. A photo diagram names the outside parts of the tank. Words in bold font in the text are defined in the glossary.

Green, Michael. *Amphibious Vehicles*. Mankato, Minn.: Capstone, 1995. 48 pp. (hc. 1-56065-219-5) $15.95.
Gr. 3–6. From the Land and Sea series. In 1938, the marines learned about an amphibious rescue vehicle built for use in the Florida Everglades. The vehicle's inventor was Donald Roebling, whose grandfather built the Brooklyn Bridge. Green chronicles the developments, improvements, and uses of amphibious vehicles. A historical map, a

photo diagram, a glossary, resources for learning more, and an index
are included.

——. *Submarines*. Mankato, Minn.: Capstone, 1998. 48 pp. (hc. 1-
    56065-555-0) $16.95.
Gr. 3–6. From the Land and Sea series. This book is a concise history
of the submarine that includes a look to the future and examines the
safety features of submarines. It contains information on inventors and
inventions that make submarines possible and ones that help to assure
the safety of the crew members. A historical map, a cutaway diagram, a
glossary, resources for learning more, and an index are provided.

Pitt, Matthew. *Apache Helicopter: The AH–64*. New York: Children's
    Press, 2000. 47 pp. (hc. 0-516-23336-X) $20.00; (pbk. 0-516-
    23536-2) $6.95.
Gr. 4–6. From the High-Tech Military Weapons series. A brief history
of helicopter credits Leonardo da Vinci with drawing the first one and
Paul Cornu with the idea of putting two rotors on helicopters. The book
describes the design and construction of the Apache helicopter. Readers
learn that the seats are made of Stephanie Kwolek's Kevlar, a bullet-
proof material. The book includes side, front, and top view diagrams. A
glossary, resources for more information, and an index are included.

——. *The Tomahawk Cruise Missile*. New York: Children's Press, 2000.
    47 pp. (hc. 0-516-23343-2) $20.00; (pbk. 0-516-23543-5) $6.95.
Gr. 4–6. From the High-Tech Military Weapons series. The book provides a brief history of the missile and explains how the Tomahawk
Cruise Missile works. Color photographs and succinct text contain a
wealth of information on these missiles providing high-interest reading
material for reluctant readers. A glossary, resources for more information, and an index are included.

Smith, Jay H. *Humvees and Other Military Vehicles*. Mankato, Minn.:
    Capstone, 1995. 48 pp. (hc. 1-56065-219-5) $15.95.
Gr. 3–6. From the Wheels series. Through four feet of water and
straight up a hill the Humvee seems unstoppable. The book provides
the specifications on the army's Humvee and then describes the Hum-
mer built for the public. The Land Rover, the Desert Patrol Vehicle,
and tanks are described in text and color photographs. A glossary, resources for learning more, and an index are included.

Staeger, Rob. *Native American Tools and Weapons*. Philadelphia: Mason Crest, 2003. 63 pp. (hc. 1-59084-132-8) $19.95.
Gr. 4–6. From the Native American Life series. The book begins with a brief account of a Nootka whale hunt, which includes descriptions of their tools. The other chapters present information on the creation of Native American tools and weapons organized geographically by regions. Illustrations and sidebars contain additional information. A chronology, a glossary, sources for further reading, and an index conclude the book.

Stein, R. Conrad. *The Manhattan Project*. Chicago: Children's Press, 1993. 32 pp. (hc. 0-316-06670-6) $21.00.
Gr. 4–7. From the Cornerstones of Freedom series. Black and white photographs, an easy to read font, and clear writing describe the Manhattan Project. The narrative ends with the pessimistic question about whether nuclear weapons will one day destroy the world. The book lacks a bibliography but does have an index.

Hansen, Ole Steen. *Military Aircraft of World War I*. New York: Crabtree, 2003. 32 pp. (hc. 0-7787-1201-X) $17.94; (pbk. 0-7787-1217-6) $8.95.
Gr. 4–8. From the Story of Flight series. This book traces the evolution of military aircraft during World War I including how better planes were designed. There is also information on the German Zeppelins, Bloody April, and the Red Baron. A plane identification guide, illustrations, a glossary, and an index are included.

——. *Military Aircraft of World War II*. New York: Crabtree, 2003. 32 pp. (hc. 0-7787-1203-6) $17.94; (pbk. 0-7787-1219-2) $8.95.
Gr. 4–8. From the Story of Flight series. Learn about the crucial role of aircraft in World War II including Pearl Harbor, the Doolittle raids, and the dropping of the atomic bombs. Illustrations, a glossary, and an index are included.

——. *Modern Military Aircraft*. New York: Crabtree, 2003. 32 pp. (hc. 0-7787-1204-4) $17.94; (pbk. 0-7787-1220-6) $8.95.
Gr. 4–8. From the Story of Flight series. From Vietnam bombers to Gulf War Stealth fighters to planes of the future, a great deal of information is packed into this book. Illustrations, a glossary, and an index are included.

O'Brien, Patrick. *Duel of the Ironclads: The Monitor vs. the Virginia.*
New York: Walker, 2003. 40 pp. (hc. 0-8027-8843-2) $18.85.
Gr. 4–8. The technology and politics behind the construction of the
ironclads is examined and their one battle during the Civil War is described. The book includes cutaway diagrams of the ironclads' interi-
ors, watercolor and gouache illustrations, maps, and an afterword.

Payan, Gregory. *Chemical and Biological Weapons: Anthrax and Sarin.* New York: Children's Press, 2000. 47 pp. (hc. 0-516-23337-8)
$20.00; (pbk. 0-516-23537-0) $6.95.
Gr. 5–8. From the High-Tech Military Weapons series. This brief overview begins with the release of sarin into the Tokyo subway. Payan
recounts the history of anthrax and sarin and describes their effects on
humans. The book ends with information on the current research and
the development of chemical weapons. A glossary, resources, and an
index are provided.

Richie, Jason. *Weapons: Designing the Tools of War.* Minneapolis: The
Oliver Press, 2000. 144 pp. (hc. 1-881508-60-9) $20.50.
Gr. 5–8. From the Innovators series. The following inventors and their
weapons are profiled: David Bushnell's submarine; Samuel Colt's re-
volver, John Ericsson's battleship; Robert Whitehead's torpedo; Hiram
Maxim's automatic machine gun; Ernest Swinton's tank; and Walter
Dornberger and Werner von Braun's ballistic missile. Photographs,
diagrams, and drawings enhance the text. A glossary, a bibliography,
and an index are included.

Meltzer, Milton. *Weapons and Warfare: From the Stone Age to the
Space Age.* Illus. by Sergio Martinez. New York: HarperCollins,
1996. 85 pp. (hc. 0-06-024875-0) $16.95.
Gr. 6–12. Meltzer selected weapons and warfare of interest to him to
include in this slim volume. He describes how the weapons came about,
why they were used, and responses to them. Technical aspects on the
weapons are not included. Black and white pencil drawings add interest
to the text. The book includes a bibliography and an index.

Tunis, Edward. *Weapons: A Pictorial History.* Baltimore, Md.: Johns
Hopkins University Press, 1999. 151 pp. (pbk. 0-8018-6229-9)
$20.95.
Gr. 6–12. Tunis offers a detailed account of the invention and im-
provement of weapons from the Stone Age to the present that is as in-

teresting today as it was when first published in 1954. Historical facts combined with anecdotes and detailed illustrations describe the evolution of weapons through the ages.

Cohen, Daniel. *The Manhattan Project.* Brookfield, Conn.: The Millbrook Press, 1999. 128 pp. (hc. 0-7613-0359-6) $24.90.
Gr. 7–10. This is a very readable history of the construction of the atomic bomb. Black and white photographs, quotes, and anecdotes add a personal touch to this narrative. A chronology, notes, a bibliography, and an index conclude the book.

Levine, Herbert M. *Chemical and Biological Weapons in Our Times.* New York: Franklin Watts, 2000. 127 pp. (hc. 0-531-11852-5) $23.00.
Gr. 8–12. Opening with the sarin nerve gas attack in the Tokyo subway in 1995, this book makes readers aware that innocent people can be randomly victimized by terrorists. This is a well-researched examination of the history of chemical and biological warfare that includes a look into the future. A glossary, sources for further reference, a bibliography, and an index are provided.

Visard, Frank, and Phil Scott, eds. *Twenty-First Century Soldier: The Weaponry, Gear, and Technology of the Military in the New Century.* New York: Time, 2002. 176 pp. (hc. 1-9319-3316-2) $24.95.
Gr. 8–12. The editors of *Popular Science* examine the aircraft, armor, ships, weaponry, gear, and tactics currently under development for the military. This look into the future is grounded in the past and explains how some proven weaponry has been transformed by integrating new technologies. An index is included.

Breuer, William B. *Secret Weapons of World War II.* New York: John Wiley and Sons, 2000. 242 pp. (hc. 0-471-37287-0) $24.95.
Gr. 10–12. This is a collection of riveting military tales focusing on the inventors, scientists, mathematicians, code makers, and the code breakers whose efforts helped to win World War II. Notes, sources, and an index conclude the book.

Buderi, Robert. *The Invention That Changed the World: How a Small Group of Radar Pioneers Won the Second World War and Launched a Technological Revolution.* New York: Simon & Schuster, 1996. 575 pp. (hc. 0-684-81021-2) $30.00.

Gr. 10–12. Readers familiar with the technological advances and the inventions resulting from space exploration may be surprised to discover that technological advances and inventions also resulted from joint British and American radar advancements during World War II. This is a riveting account of the work of the radar pioneers during the war and afterwards. An epilogue, notes, a glossary, a list of interviews, a bibliography, and an index are all found at the end of the book.

Conant, Jennet. *Tuxedo Park: A Wall Street Tycoon and the Secret Palace of Science That Changed the Course of World War II.* New York: Simon & Schuster, 2002. 331 pp. (hc. 0-684-87287-0) $26.00.
Gr. 10–12. Alfred L. Loomis's home in Tuxedo Park, N.Y. became the gathering place for scientists including Einstein, Bohr, and Fermi. Their critical research led to the development of radar systems that hastened the end of World War II. An epilogue, a list of Loomis's scientific publications, notes on sources, and an index are included.

Fermi, Rachel, and Esther Samra. *Picture the Bomb: Photographs from the Secret World of the Manhattan Project.* New York: Harry N. Abrams, 1995. 232 pp. (hc. 0-8109-3735-2) $95.00.
Gr. 10–12. This is a photographic essay of the Manhattan Project. Richard Rhodes, Pulitzer Prize–winning author of *The Making of the Atomic Bomb*, wrote the introduction and provides a historical outline for the project. Rachel Fermi is the granddaughter of Enrico Fermi and was able to gain access to personal and family archives of many of the participants involved in the Manhattan Project. An afterword, biographies and profiles, a timeline, a glossary, an annotated bibliography, and an index conclude the book.

McDaid, Hugh, and David Oliver. *Smart Weapons: Top Secret History of Remote Controlled Airborne Weapons.* New York: Welcome Rain, 1997. 208 pp. (pbk. 1-56649-287-4) $29.95.
Gr. 10–12. This is the story of Unmanned Aerial Vehicles (UAVs) that relay military intelligence. As the technology advances so do UAVs' roles in warfare. Photographs of UAVs and illustrations of future UAVs grab the reader's attention. Fact files, Internet sites, an index, and abbreviations conclude the book.

O'Connell, Robert L. *Soul of the Sword: An Illustrated History of Weaponry and Warfare from Prehistory to the Present.* Illus. by

John Batchelor. New York: The Free Press, 2002. 390 pp. (hc. 0-684-84407-9) $35.00.

Gr. 10–12. O'Connell combines technology, sociology, and economics to portray the history of weaponry and its impact on society. Illustrations of the weapons, illustrations of the combatants, and sidebars describing specific weaponry are interspersed in this informative narrative. The book includes an epilogue, notes, a selected bibliography, and an index.

Broad, William J. *Teller's War: The Top-Secret Story Behind the Star Wars Deception.* New York: Simon & Schuster, 1992. 350 pp. (hc. 0-671-70106-1) $25.00.

Gr. 11–12. Two-time Pulitzer Prize–winner Broad takes readers behind the scenes to examine the politics behind the Strategic Defense Initiative–Star Wars. This experimental x-ray laser project is described and so are its flaws. Some of the concepts and ideas presented in the book are explained in black and white line drawings and diagrams. The book concludes with an epilogue, an extensive collection of notes, a long bibliography, and an index.

Miller, Judith, Stephen Engelberg, and William Broad. *Germs: Biological Weapons and America's Secret War.* New York: Simon & Schuster, 2001. 382 pp. (hc. 0-684-87158-0) $27.00.

Gr. 11–12. This examination of germ warfare begins with the riveting story of a salmonella attack on an Oregon community conducted by cultists. The history of germ warfare is recounted and possible future scenarios are discussed. This is provocative material sure to form a basis for lively classroom discussions. Conclusions, notes, a selected bibliography, and an index are provided.

Chapter 14

Invention Collections

Books in this chapter contain information on collections of inventions that encompass a variety of areas such as transportation, everyday life, and communications. Books on great inventions, inventions that happened by mistake, inventions that changed the world, and inventions from different time periods are some of the intriguing topics found in this chapter.

Taylor, Barbara. *I Wonder Why Zippers Have Teeth and Other Questions about Inventions*. New York: Kingfisher, 1996. 32 pp. (hc. 1-85697-670-X) $9.56; (pbk. 1-85697-688-2) $9.90.
Gr. K–3. Short, simple questions about inventions are answered with brief responses illustrated by a mixture of cartoons and realistic illustrations. Throughout the text, inventors and their inventions are introduced. An index is included.

Harper, Charise Mericle. *Imaginative Inventions: The Who, What, Where, When and Why of Roller Skates, Potato Chips, Marbles, and Pie*. Boston: Little, Brown, 2001. 32 pp. (hc. 0-316-34725-6) $14.95.
Gr. 1–4. Fact and fancy combined with, at times, awkward verses and colorful, whimsical drawings across two-page spreads tell the origins of things of interest to children including potato chips, Frisbees, chewing gum, marbles, and animal cookies. By turning the book sideways students can read additional information about the invention written down the side of the page.

Crawford, Jean B., and Karin Kinney, eds. *Inventions*. Richmond, Va.:
    Time-Life Books, 1999. 88 pp. (hc. 0-8094-9454-X) $14.95.
Gr. 2–5. From A Child's First Library of Learning series. The book
uses a question and answer format accompanied by cartoon-like illustrations to describe inventions and their origins. "To the Parent" boxes
contain additional information. The last section of the book contains
workbook activities to review and extend the material.

Goldberg, Rube. *The Best of Rube Goldberg*. Upper Saddle River, N.J.:
    Prentice Hall, 1979. 130 pp. (hc. 0-13-074807-2) $9.95.
Gr. 3–6. This is wacky collection of labor-saving devices complete
with step-by-step instructions on how they work and illustrations. This
book gets young imaginations going and encourages them to create
their own convoluted devices.

Hughes, Susan. *Canada Invents!* Toronto, Ont.: Maple Tree Press,
    2002. 112 pp. (hc. 1-894379-23-3) $29.95; (pbk. 1-894379-24-1)
    $19.95.
Gr. 3–6. From the Wow Canada! series. This collection of Canadian
inventions is organized into themes including coping with the cold,
energy and power, communications, transportation, and physical challenges. While the focus of the book is on inventions, it also has brief
biographies of inventors. The book contains photographs, illustrations,
and an index.

Parker, Steve. *Fifty-Three and a Half Things that Changed the World
    and Some That Didn't*. Brookfield, Conn.: The Millbrook Press,
    1992. 62 pp. (hc. 1-56294-603-X) $23.90; (pbk. 1-56294-894-6)
    $8.95.
Gr. 3–6. The book is divided into three sections: inventions from ancient societies; inventions that changed people's lives; and inventions
that are useful, but did not change people's lives. Inventions that
changed lives include such things as the clock, the telescope, the electric motor, and the television. Inventions that did not change lives include the safety pin, the rocking chair, the teddy bear, and invisible ink.
An index is included.

Romanek, Trudee. *The Technology Book for Girls and Other Advanced
    Beings*. Illus. by Pat Cupples. Tonawanda, N.Y.: Kids Can Press,
    2001. 56 pp. (hc. 1-55074-936-6) $14.95; (pbk. 1-55074-619-7)
    $8.95.

Gr. 3–6. Each chapter introduces an everyday object with a conversation between school friends or students and family members. Then, colorful illustrations, sidebars, and easy-to-understand text explain the object and how it works. The brief chapters conclude with experiments for learning more. Ideas for science fair projects, a note to adults, and an index are provided.

Gates, Phil. *Wild Technology: Inventions Inspired by Nature.* New York: Kingfisher, 1999. 79 pp. (pbk. 0-7534-5261-8) $8.95.
Gr. 3–8. This book demonstrates how many inventions are modeled after nature. With catchy titles like "squid to jet engine" and "tooth to chisel," the two-page spreads entice readers to read carefully and examine the illustrations. Color photographs, cutaway views, diagrams, sidebars, and illustrations add to the clear, succinct text. A glossary and an index are included.

Jones, Charlotte Foltz. *Accidents May Happen: Fifty Inventions Discovered by Mistake.* Illus. by John O'Brien. New York: Dell, 1996. 86 pp. (hc. 0-385-32162-7) $16.95.
Gr. 3–8. Here is a collection of brief, intriguing entries describing how different inventions were discovered, some by accident. Black and white cartoon-like illustrations complement the text. The entries in the book are grouped by categories, such as "Fed Up," "Patriotic Accidents," and "Explosive Discoveries." The book concludes with resources, a bibliography, and an index.

Wilcox, Jane. *Why Do We Use That?* Danbury, Conn.: Franklin Watts, 1996. 32 pp. (hc. 0-531-14395-3) $20.00.
Gr. 3–8. From the Why Do We series. Small, captioned, cartoon-like illustrations tell a brief history of a variety of inventions. This is a book for browsing and one that students pick up more than once. A quiz and an index conclude the book.

Crosher, Judith. *Technology in the Time of Ancient Egypt.* Austin, Tex.: Raintree Steck-Vaughn, 1998. 48 pp. (hc. 0-8172-4875-7) $27.12.
Gr. 3–9. From Technology in the Time of series. This well-illustrated book gives a thumbnail sketch of life in early Egypt chronicling both discoveries and inventions of the period. Inventions include a wooden plow, wooden sickles, a wild duck trap, a wine press, a shaduf, and many more. The arts of baking and of making cloth are also explained.

The book includes a timeline, a glossary, resources containing further information, and an index.

——. *Technology in the Time of Ancient Greece.* Austin, Tex.: Raintree Steck-Vaughn, 1998. 48 pp. (hc. 0-8172-4877-3) $27.12.
Gr. 3–9. From Technology in the Time of series. This book not only describes discoveries and inventions from ancient Greece, it also includes hands-on activities for students to try, such as making jewelry. Colorful diagrams explain how some of the technology worked. A timeline, a glossary, books to read, and an index are included.

——. *Technology in the Time of the Maya.* Austin, Tex.: Raintree Steck-Vaughn, 1998. 48 pp. (hc. 0-8172-4881-1) $27.12.
Gr. 3–9. From Technology in the Time of series. Diagrams, photographs, illustrations, recipes, instructions for games and crafts, and short paragraphs of text explain the technology used and developed by the Maya. A timeline, a glossary, books to read, and an index are included.

Hicks, Peter. *Technology in the Time of the Vikings.* Austin, Tex.: Raintree Steck-Vaughn, 1998. 48 pp. (hc 0-8172-4880-3) $27.12.
Gr. 3–9. From Technology in the Time of series. Readers learn a great deal about the Vikings as they examine the information in this book on Viking technology. Colorful illustrations and diagrams accompany the succinct text. A timeline, a glossary, books to read, and an index are included.

Morgan, Nina. *Technology in the Time of the Aztecs.* Austin, Tex.: Raintree Steck-Vaughn, 1998. 48 pp. (hc. 0-8172-4878-1) $27.12.
Gr. 3–9. From Technology in the Time of series. This book provides readers a brief overview of the Aztec culture as it examines the technology they developed and used. A timeline, a glossary, books to read, and an index are included.

Snedden, Robert. *Technology in the Time of Ancient Rome.* Austin, Tex.: Raintree Steck-Vaughn, 1998. 48 pp. (hc. 0-8172-4876-5) $27.12.
Gr. 3–9. From Technology in the Time of series. Descriptions of discoveries and inventions in early Rome include ones for producing food, making clothing, constructing buildings, and supplying water. Illustra-

tions enhance the descriptions. There is a timeline, a glossary, a list for further reading, and an index.

Jones, Charlotte Foltz. *Mistakes that Worked.* Illus. by John O'Brien. New York: Doubleday, 1991. 82 pp. (hc. 0-385-32043-4) $11.95.
Gr. 4–6. Brown'n' serve rolls were invented when a volunteer fireman, Joe Gregor, took his undercooked rolls out of the oven before rushing off to a fire. When he returned, he put them back in the oven to finish cooking. Ivory soap was invented when a worker went to lunch and left a batch of soap mixing for longer than normal. These invention stories and others are told with humor and colorful cartoon drawings. An index is included.

Williams, Brian, and Brenda Williams, eds. *World Book Looks at Inventions and Discoveries.* Chicago: World Book, 1996. 63 pp. (pbk. 0-7166-1802-8) $6.95.
Gr. 4–8. From the World Book Looks at series. This compact examination of inventions and discoveries begins with information on why people invent and how people invent. The inventions and discoveries highlighted in the book include those found from factories to farms and from the earth to space. Short paragraphs, sidebars, photographs, reproductions, and diagrams tell the story of inventions through the ages. A timeline, a glossary, and an index conclude the book.

Wood, Richard, ed. *Great Inventions.* Richmond, Va.: Time-Life Books, 1995. 64 pp. (hc. 0-7835-4766-8) $16.00.
Gr. 4–8. From the Nature Company Discoveries Library series. Photographs, diagrams, sidebars, and short chunks of text in a collage format introduce inventions from everyday life; transportation; communication; instruments and machines; power and energy; war and peace; and life and medicine. A glossary and an index conclude the book.

Wulffson, Don L. *The Kid Who Invented the Popsicle: And Other Surprising Stories about Inventions.* New York: Dutton Children's Books, 1999. 114 pp. (hc. 0-525-65221-3) $13.99; (pbk. 0-141-30204-6) $4.99.
Gr. 4–8. This collection of stories is about inventions of interest to children, such as robots, Tinker Toys, ice cream cones, and jigsaw puzzles. The inventions are listed alphabetically and each is allotted a brief one-page description. These entries are fun to read aloud to children.

——. *The Kid Who Invented the Trampoline: More Surprising Stories about Inventions*. New York: Dutton Children's Books, 2001. 120 pp. (hc. 0-525-46654-1) $15.99.
Gr. 4–8. This is the sequel to *The Kid Who Invented the Popsicle*. Readers learn how and why fifty inventions ranging from animated cartoons to the Zamboni were created. Illustrations and photographs accompany the stories as well as small text boxes filled with tantalizing tidbits of information.

——. *Toys!: Amazing Stories behind Some Great Inventions*. Illus. by Laurie Keller. New York: Henry Holt, 2000. 137 pp. (hc. 0-8050-6196-7) $15.95.
Gr. 4–8. This is a collection of captivating stories about some of the best-loved toys such as Lego, Mr. Potato Head, the Slinky, and Twister. Cartoon-like illustrations are interspersed throughout the text. The history of each toy's invention is enhanced by the accompanying trivia. The book concludes with a bibliography and websites.

Yenne, Bill. *100 Inventions that Shaped World History*. San Francisco: Bluewood Books, 1993. 112 pp. (pbk. 0-912517-02-6) $47.95.
Gr. 4–8. Brief descriptions and black and white illustrations describe an intriguing collection of inventions ranging from aspirin to virtual reality. The inventions are arranged chronologically. The book concludes with a trivia quiz, suggested projects, and an index.

Lasson, Kenneth. *Mousetraps and Muffling Cups: One Hundred Brilliant and Bizarre U.S. Patents*. New York: Arbor House, 1986. 219 pp. (pbk. 0-87795-786-X) $9.95.
Gr. 4–12. From air brakes to the dimple maker readers browse through an amazing collection of inventions. On the left side of the two-page spread is the name of the invention, the patent number, the name of the inventor, and a brief description of the invention. On the opposite page is a black and white line drawing of the invention. This is a fun book to examine and sure to give students ideas for their own inventions.

Murphy, Jim. *Guess Again: More Weird and Wacky Inventions*. New York: Bradbury, 1986. 91 pp. (hc. 0-02-767720-6) $12.95.
Gr. 4–12. From the early days of this country, inventions were creative and innovative items to aid people in protecting themselves (the rifle) or making life easier (the Johnny Lift) or perhaps saving themselves (the coffin with an escape hatch). This delightful book will cause the

reader to be surprised with the ingenuity of inventors and to laugh at their wacky ideas. The chapters include animals, clothing, mobility, fun and games, personal hygiene, odds-and-ends, and a handful of stories. An afterword and resources for learning more are included.

———. *Weird and Wacky Inventions*. New York: Crown Publishers, 1978. 92 pp. (hc. 0-517-53318-9) $12.95.
Gr. 4–12. Read the question about the invention, examine the black and white line drawing of the invention, and then pick the correct multiple-choice answer to the question. Turn the page to find the answer to the question and an explanation of how the invention worked. Expect the unexpected, these are weird and wacky inventions. After the invention questions Murphy shows how the velocipede and the bellows evolved. An afterword discusses patents.

Parker, Steve. *1900–20: A Shrinking World*. Milwaukee: Gareth Stevens, 2000. 32 pp. (hc. 0-8368-2942-5) $23.95.
Gr. 5–8. From the Twentieth Century Science and Technology series. Exploring the atom, flying, motion pictures, automobiles, the discovery of oil, and medical advances such as x-rays mark this period in history. Succinct descriptions, color photographs, and labeled diagrams fill the pages. The book concludes with a timeline, a glossary, more books to read, websites, and an index.

———. *1920–40: Atoms to Automation*. Milwaukee: Gareth Stevens, 2000. 32 pp. (hc. 0-8368-2943-3) $23.95.
Gr. 5–8. From the Twentieth Century Science and Technology series. Looking into deep space, developing liquid fuel rockets, improving transportation, and developing of modern materials such as Bakelite and rayon were technological advances during this period. Clear, colorful photographs and diagrams enhance the text. The book concludes with a timeline, a glossary, more books to read, websites, and an index.

———. *1940–60: The Nuclear Age*. Milwaukee: Gareth Stevens, 2000. 32 pp. (hc. 0-8368-2944-1) $23.95.
Gr. 5–8. From the Twentieth Century Science and Technology series. Unraveling the secrets of DNA, unleashing the power of the atom, building nuclear power plants, and developing transistors are a few of the technological innovations described in this book. Color photographs, labeled diagrams, and concise narrative explain the innovations.

The book concludes with a timeline, a glossary, more books to read, websites, and an index.

——. *1960s: Space and Time*. Milwaukee: Gareth Stevens, 2000. 32 pp. (hc. 0-8368-2945-X) $23.95.
Gr. 5–8. From the Twentieth Century Science and Technology series. While the focus of this period is the race to the moon, there is also information and pictures on modular housing, plastics, lasers, and other gadgets. The pages are filled with color photographs and labeled diagrams accompanied by succinct descriptions. The book concludes with a timeline, a glossary, more books to read, websites, and an index.

——. *1970–90: Computers and Chips*. Milwaukee: Gareth Stevens, 2001. 32 pp. (hc. 0-8368-2946-8) $23.95.
Gr. 5–8. From the Twentieth Century Science and Technology series. This was a time of extended explorations into space, environmental pollution, alternative energy solutions, and the advent of computers. The pages are filled with color photographs and labeled diagrams accompanied by succinct descriptions. The book concludes with a timeline, a glossary, more books to read, websites, and an index.

——. *1990–2000: The Electronic Age*. Milwaukee: Gareth Stevens, 2000. 32 pp. (hc. 0-8368-2947-6) $23.95.
Gr. 5–8. From the Twentieth Century Science and Technology series. Virtual reality, joint science ventures, greater concern for the environment, and the microchip revolution marked this period. Photographs and diagrams complement the text. The book concludes with a timeline, a glossary, more books to read, websites, and an index.

Thimmesh, Catherine. *Girls Think of Everything: Stories of Ingenious Inventions by Women*. Illus. by Melissa Sweet. Boston: Houghton Mifflin, 2000. 59 pp. (hc. 0-395-93744-2) $16.00; (pbk. 0-618-19563-7) $6.95.
Gr. 5–9. The book begins and ends with a timeline of women's inventions ranging from 3000 B.C. to 1995. Included are the stories of selected women inventors who responded to everyday problems with their own unique solutions. For example, Mary Anderson's windshield wipers, Stephanie Kwolek's Kevlar for bullet proof vests, and ten-year-old Becky Schroeder's Glo-sheet for writing in the dark. The book concludes with resources for young inventors, books for further reading, a list of sources, and an index.

——. *The Sky's the Limit*. Illus. by Melissa Sweet. Boston: Houghton Mifflin, 2002. 73 pp. (hc. 0-618-07698-0) $16.00.
Gr. 5–9. Curious females discover and invent, but are often ignored in the history books. Timmesh introduces readers to successful women and young girls who let their curiosity lead them to wondrous discoveries. The book includes a collection of resources for further exploration, a selected timeline of discoveries by women, and an index.

Tomecek, Stephen M. *What a Great Idea! Inventions that Changed the World*. Illus. by Dan Stuckenschneider. New York: Scholastic, 2003. 112 pp. (hc. 0-590-68144-3) $18.95.
Gr. 5–12. This book describes and illustrates forty-five technological touchstones from the hand ax and mathematics to computer chips and the laser. The format of the book includes the problem to be solved and information on who solved the problem. This is followed by a description of how the invention or discovery works, the impact of it on society, and then a discussion of other inventions or discoveries that came from the original one. A bibliography, websites, invention contests, and an index conclude the book.

Bridgman, Roger. *1000 Inventions and Discoveries*. New York: DK Publishing, 2002. 256 pp. (hc. 0-7894-8826-4) $24.99.
Gr. 7–12. A thin timeline runs across the bottom of the pages beginning with the year 1,600,000 B.C. and ending with the year 2001. Photographs, drawings, reproductions, and brief descriptions tell the stories of a multitude of inventions. The book includes an index of inventions and discoveries as well as an index of inventors and discoverers.

Dyson, James. *A History of Great Inventions*. Emeryville, Calif.: Avalon, 2001. 192 pp. (hc. 0-7628-6683-7) $32.00.
Gr. 7–12. This book is more than a history of inventions as Dyson also examines the situations and urges behind the inventions. He also chronicles how frustration, serendipity, perseverance, and hard work are part of the invention process. Illustrations and an index are included.

Flatow, Ira. *They All Laughed . . . From Light Bulbs to Lasers: The Fascinating Stories Behind the Great Inventions That Have Changed Our Lives*. New York: HarperCollins, 1992. 240 pp. (hc. 0-06-016445-X) $20.00.

Gr. 7–12. Here are the very human stories behind inventions we use everyday. With chapter titles like "The Blender That Conquered Polio" and "The Wasp That Changed the World" readers know they are in for a delightful tour of the world of inventing. An epilogue, source notes, books for further reading, and an index are included.

Kassinger, Ruth. *Reinvent the Wheel: Make Classic Inventions, Discover Your Problem-Solving Genius, and Take the Inventor's Challenge*. Hoboken, N.J.: John Wiley and Sons, 2001. 128 pp. (pbk. 0-471-39539-0) $14.95.
Gr. 7–12. There are four parts to the book including inventions based on simple machines, inventions involving chemical changes, inventions for predicting the weather, and inventions that use magnetism and electricity. This is a collection of hands-on projects that show students how to recreate simple inventions and that challenge them to design and create their own inventions. A glossary and an index are included.

Wolfe, Maynard Frank, and Rube Goldberg. *Rube Goldberg: Inventions!* New York: Simon & Schuster, 2000. 192 pp. (hc. 0-684-86724-9) $25.00.
Gr. 7–12. Meet cartoonist Rueben Lucius Goldberg and his drawings of intricate, fanciful inventions in this amusing book. Inventions from the Rube Goldberg Machine Contest held at Purdue University each year are included.

Bud, Robert, Simon Niziol, Timothy Book, and Andrew Nahum. *Inventing the Modern World: Technology Since 1750*. New York: DK Publishing, 2000. 224 pp. (hc. 0-7894-6828-X) $25.00.
Gr. 8–12. Beginning with Great Britain's Industrial Revolution in the eighteenth century, this book covers two hundred fifty years of technological change. Photographs and illustrations fill the pages of this concise text. Suggestions for further reading and an index conclude the text.

Macdonald, Anne L. *Feminine Ingenuity: Women and Invention in America*. New York: Ballantine Books, 1992. 514 pp. (hc. 0-345-35811-2) $22.50.
Gr. 8–12. Researching the archives of the patent office provided only part of the story of the importance of women's inventions to America's progress. Macdonald also searched manuscripts, magazines, lectures, and the records of major fairs and expositions to document women's

contributions. It was not until 1809 that a woman was granted a patent in her own name; prior to that, women's inventions were patented in men's names. The obstacles women had to overcome to have their inventions recorded in the patent office are placed in a historical context. An appendix lists women inventors and their patents that are noted in the book. A notes section, a bibliography and an index are included.

van Dulken, Stephen. *Inventing the Twentieth Century: 100 Inventions that Shaped the World: From the Airplane to the Zipper.* Washington Square, N.Y.: New York University Press, 2000. 246 pp. (hc. 0-8147-8808-4) $30.00; (pbk. 0-8147-8812-2) $16.95.
Gr. 8–12. These inventions were drawn from the patent collection of the British Library. Entries include a brief history of the patent and a copy of the original patent application. The entries are grouped by decade and each section begins with a brief introduction to world events, which provides a context for understanding the impact of the inventions. The book concludes with suggestions for further reading, and an index.

Bragg, Melvyn, with Ruth Gardiner. *On Giants' Shoulders: Great Scientists and Their Discoveries from Archimedes.* New York: John Wiley and Sons, 1998. 365 pp. (hc. 0-471-15732-4) $22.95.
Gr. 9–12. This collection of essays is based on discussions about scientists and their discoveries. These discussions with scientists and historians took place on Bragg's radio program. A bibliography is included.

Brockman, John, ed. *The Greatest Inventions of the Past Two Thousand Years.* New York: Simon & Schuster, 2000. 192 pp. (hc. 0-684-85998-X) $22.00.
Gr. 9–12. A poll was taken soliciting the most important inventions through the year 2000. Brockman researched the information obtained and edited this book of the results. Candidates for the greatest invention include the obvious (computers) to the less obvious (the Indo-Arab counting system). Some of the "inventions" are actually concepts.

Burke, James. *Circles: Fifty Round Trips through History, Technology, Science, Culture.* Illus. by Dusan Petricic. New York: Simon & Schuster, 2000. 286 pp. (hc. 0-7432-0008-X) $24.00.
Gr. 9–12. Most people (especially historians) would agree that learning about the past helps in predicting the future. The author further believes

that some inventions and discoveries circle back to the original. Fifty circular tales taken from his *Scientific American* column recount how interconnectedness and serendipity are part of the process of discovery and invention. These are fascinating vignettes that twist and turn and end up where they began. A select bibliography concludes the book.

Gratzer, Walter. *Eurekas and Euphoria: The Oxford Book of Scientific Anecdotes.* New York: Oxford University Press, 2002. 301 pp. (hc. 0-19-280403-0) $28.00.
Gr. 9–12. One hundred eighty-one anecdotes of discovery portray the personalities and idiosyncrasies of familiar scientists. Gratzer begins by stating that anecdotes must be both entertaining and thought-provoking and indeed the ones he recounts fit both of these criteria. Name and subject indexes are included.

Gross, Ernie. *Advances and Innovations in American Daily Life, 1600–1930.* Jefferson, N.C.: McFarland, 2002. 298 pp. (pbk. 0-7864-1248-8) $35.00.
Gr. 9–12. Sections in the book include agriculture, art, music, business, finance, clothing, communications, education, energy, entertainment, food, health, labor, law, manufacturing, public service, religion, science, shelter, domestic furnishings, social welfare, sports, and transportation. This is a concise examination of familiar innovations. An index is included, but the book lacks references.

Hillman, David, and David Gibbs. *Century Makers: One Hundred Clever Things We Take for Granted Which Have Changed Our Lives Over the Last One Hundred Years.* New York: Welcome Rain, 1999. 191 pp. (pbk. 1-56649-000-6) $16.95.
Gr. 9–12. While we are accustomed to reading and learning about major inventions such as electricity and airplanes, there are countless smaller inventions, which have changed our lives, such as paperclips, Velcro, and washing machines. This book gives short explanations of these inventions and their inventors. The book is written in a light-hearted way, which will entertain as well as inform readers.

Vare, Ethlie Ann, and Greg Ptacek. *Mothers of Invention: From the Bra to the Bomb, Forgotten Women and Their Unforgettable Ideas.* New York: William Morrow, 1988. 220 pp. (hc. 0-688-06464-7) $17.95.

Gr. 9–12. In the face of gender discrimination women inventors persist. This book is a tribute to these persistent, hardworking women inventors and an inspiration to young female inventors. These short, engaging narratives help give these inventors the recognition they deserve. Appendixes and an index are included.

——. *Patently Female: From AZT to TV Dinners: Stories of Women Inventors and Their Breakthrough Ideas*. New York: John Wiley and Sons, 2002. 220 pp. (hc. 0-471-02334-5) $24.95.

Gr. 9–12. While most people can easily name several male inventors, they are often unable to name female inventors. The authors contend that this is because women's inventions are not written about and this book, an informal sequel to their book *Mothers of Invention: From the Bra to the Bomb, Forgotten Women and Their Unforgettable Ideas*, seeks to correct the oversight. Women's inventions and discoveries include Scotchgard, bone marrow transplants, the Mars Rover, and computer languages. A timeline, additional resources, and an index are included.

Burke, James. *The Pinball Effect: How Renaissance Water Gardens Made the Carburetor Possible—and Other Journeys through Knowledge*. Boston: Little, Brown, 1996. 310 pp. (hc. 0-316-11602-5) $23.95; (pbk. 0-316-11610-6) $15.95.

Gr. 10–12. Just as a pinball bounces and careens from place to place, so too do the processes that led us to change and innovation. Burke takes readers on a wondrous ride from invention to invention, along the way encountering serendipitous discoveries that led to further innovations. The margins of the book contain cross-references to other parts of the book much the same as a hypertext document, which allows readers to read the text in a nonlinear fashion. A select bibliography and an index are included.

Feldman, Burton. *The Nobel Prize: A History of Genius, Controversy, and Prestige*. New York: Arcade, 2000. 489 pp. (hc. 1-59970-537-X) $29.95.

Gr. 10–12. Beginning with the work and bequest of Alfred Nobel this book provides a readable, informative history of the Nobel Prize including blunders and controversies surrounding the prize. Appendixes in the book contain lists of the value of prizes, prizes by nation, women laureates, family laureates, and Jewish laureates. The book includes a chronology of prizes, notes, a selected bibliography, and an index.

Frugoni, Chiara. *Books, Banks, Buttons: And Other Inventions from the Middle Ages*. Trans. by William McCuaig. New York: Columbia University Press, 2001. 178 pp. (hc. 0-231-12812-6) $27.95.

Gr. 10–12. Eyeglasses, forks, and wheelbarrows all came from the Middle Ages. Illustrations include paintings from the Middle Ages that show the inventions in use. Quotes, including some from the Bible, tell about the inventions. Notes, a select bibliography, and an index of persons are included. The book does not include an index of inventions.

Chapter 15

How Things Work

Nonfiction books in this chapter explore the inner workings of a wide variety of inventions in order to explain how things work. Colorful illustrations and detailed diagrams help readers understand the technology behind familiar objects. These books answer the questions posed by curious minds.

What's Inside Toys? New York: Dorling Kindersley, 1991. 17 pp. (hc. 1-879431-08-4) $8.95.
Gr. P–1. Young children who wonder how a baby doll cries tears can find the answer in this book. Color pictures and labeled cutaway diagrams show children what is inside eight favorite toys and how they work.

Porter, Alison, and Eryl Davies, eds. *How Things Work.* New York: Barnes and Noble Books, 2003. 64 pp. (hc. 0-7607-4642-7) $6.95.
Gr. K–3. The book begins by explaining solar power, wind power, and how these two forces produce electricity. It includes sections on machines; communications; leisure and entertainment; and transportation. Colorful photographs, cutaway diagrams, and illustrations help to explain how things work. At the end is an explanation of some scientific principles, a glossary, and an index.

Graham, Ian. *Boats, Ships, Submarines, and Other Floating Machines.* New York: Kingfisher Books, 1993. 40 pp. (pbk. 1-85697-867-2) $13.90.
Gr. 3–6. From the How Things Work series. An introduction explains why some objects float and other objects sink. From there readers are

introduced to a variety of seagoing crafts and Graham explains what makes them float. The book includes illustrations, diagrams, cutaway views, hands-on activities, and an index.

Butterfield, Moira. *Look Inside Cross-Sections: Record Breakers and Other Speed Machines*. Illus. by Chris Grigg and Keith Harmer. New York: Dorling Kindersley, 1995. 32 pp. (pbk. 0-7894-0320-X) $5.95.
Gr. 3–8. Exploded views and diagrams explain how these record-breaking machines run and describe the technology that makes them possible. These machines set records on racetracks, in the water, in the air, and on railroad tracks. An illustrated glossary and an index are included.

Ardley, Neil. *How Things Work*. Pleasantville, N.Y.: Reader's Digest, 1995. 191 pp. (hc. 0-89577-694-4) $24.00.
Gr. 4–8. Readers discover how things work through brief descriptive text, diagrams, color photographs, and hands-on activities. The book begins with the basics of machines and then the chapters are divided into construction and buildings, household machines, transportation, leisure, and information technology. A glossary and an index are provided.

Bartholomew, Alan. *Electric Gadgets and Gizmos: Battery-Powered Buildable Gadgets that Go*. Illus. by Lynn Bartholomew. Buffalo, N.Y.: Kids Can Press, 1998. 47 pp. (pbk. 1-55074-439-9) $5.95.
Gr. 4–8. The book begins with information on the materials needed to make the gadgets and gizmos in the book, how to make battery connections, and how to make switches. The step-by-step instructions for making gadgets such as a flashlight, a communication buzzer, and a squirt finger are accompanied by realistic illustrations.

Farndon, John. *What Happens When . . . ?* Illus. by Steve Fricker and Mike Harnden. New York: Scholastic, 1996. 45 pp. (hc. 0-590-84754-6) $17.95.
Gr. 4–8. A multitude of little people take readers on a fact-filled, action-packed journey of discovery as they explain in pictures and words thirteen processes including how a newspaper is made, how sewage is treated, and how mail moves from place to place. A glossary and an index conclude the book.

Challoner, Jack. *Flight*. Illus. by Jon Barnes. New York: Thomson
    Learning, 1995. 48 pp. (hc. 1-56847-347-8) $12.95; (pbk. 1-58728-
    355-7) $6.95.
Gr. 5–8. From the Make It Work! Science series. This is a collection of
hands-on projects that involve children in making a variety of flying
models to demonstrate the properties of flight. Some of the projects
require adult assistance and supervision. A glossary and an index conclude the book.

Senior, Kathryn. *Photography*. Illus. by Jon Barnes. New York: World
    Book, 1996. 48 pp. (hc. 1-7166-1731-5) $6.95.
Gr. 5–8. From the Make It Work! Science series. With step-by-step
instructions and color photographs Senior explains how to build simple
cameras, take photographs, and develop them. Some of the projects are
complex and require supplies from a photographic shop. A glossary and
an index conclude the book.

Macaulay, David, with Neil Ardley. *The New Way Things Work*. Bos-
    ton: Houghton Mifflin, 1998. 400 pp. (hc. 0-395-93847-3) $35.00.
Gr. 5–12. Detailed drawings, labeled cross sections, succinct technical
writing, and an amusing wooly mammoth thrown in for good measure
combine to make this a book to browse or to use as a comprehensive
guide to how things work. This is an updated version of the book published in 1988 and it contains technological innovations created since
the original publication. A glossary of technical terms and an index are
provided.

Pinchuk, Amy. *Make Amazing Toy and Game Gadgets*. Illus. by Allan
    Moon and Tina Holdcroft. New York: HarperCollins, 2001. 64 pp.
    (hc. 0-688-17797-2) $16.95.
Gr. 6–12. From the Popular Mechanics for Kids series. What better
way to learn how things work than to make some toys and gadgets? As
students construct the toys and gadgets, they learn how they operate.
Detailed, illustrated instructions are included for a light box, flashy
jewelry, a camera, a buzz-off game, and sunglasses. Safety rules, how-
to tips, a glossary, and an index are included.

Suplee, Curt. *Everyday Science Explained*. Washington, D.C.: National
    Geographic Society, 1996. 272 pp. (pbk. 0-7922-7194-7) $19.95.
Gr. 6–12. The physics, chemistry, biology, and biochemistry encountered in everyday life are explained in this fascinating book. Photo-

graphs, cartoons, cutaway diagrams, cross-references, and sidebars enhance the text by providing additional resources for understanding the concepts. An index and a bibliography are included.

The Ultimate Book of Cross Sections. New York: DK Publishing, 1996. 304 pp. (hc. 0-7894-1195-4) $45.00.
Gr. 6–12. The colorful cross sections are spread across two pages. They include a textbox of technical data and information on the technology that operates the machine. Sections include cars, trains, bulldozers, tanks, rescue vehicles, record breakers, ships, planes, jets, and space. An index is included.

Brain, Marshall. *Marshall Brain's How Stuff Works*. New York: John Wiley and Sons, 2001. 320 pp. (hc. 0-7645-6518-4) $24.99.
Gr. 9–12. Colorful illustrations and diagrams with easy-to-understand explanations show readers how things work. In the air, on the road, on the web, and around the house are just some of the locations readers explore as they discover the inner workings of a variety of objects. An index concludes the book.

——. *Marshall Brain's More How Stuff Works*. New York: John Wiley and Sons, 2002. 320 pp. (hc. 0-7645-6711-X) $24.99.
Gr. 9–12. Discover the inner workings of toys, heavy equipment, nature, office machines, the Internet, and a variety of other inventions and discoveries all around you. Detailed descriptions, labeled three-dimensional diagrams, and colorful photographs explain how things work. The book includes an index.

Brain, Marshall, ed. *Marshall Brain's How Stuff Works: How Much Does the Earth Weigh?* New York: John Wiley and Sons, 2001. 144 pp. (hc. 0-7645-6519-2) $24.99.
Gr. 9–12. The questions and answers in this book came from the HowStuffWorks.com website "Question of the Day." A variety of topics are covered in the eleven chapters such as food, health, entertainment, and special effects. An index concludes the book.

Langone, John. *National Geographic's How Things Work: Everyday Technology Explained*. Illus. by Pete Samek, Andy Christie, and Bryan Christie. Washington, D.C.: National Geographic Society, 1999. 272 pp. (hc. 0-7922-7150-5) $35.00.

Gr. 9–12. Each entry includes a brief history of the technology and its uses. Clear, bright photographs and three-dimensional drawings show the inner workings of items and processes such as the garbage disposal, photograph processing, a telephone network, and fireworks. An afterword discusses the impact of technology for better or worse on our lives. A glossary, resources for learning more, and an index are included.

White, Ron. *How Computers Work*. Illus. by Timothy Edward Downs and Stephen Adams. Indianapolis, Ind.: Macmillan Computer Publishing, 1999. 421 pp. (pbk. 0-7897-2112-0) $34.99.
Gr. 9–12. White's descriptions of how thing work are not limited to the computer; he also includes software, multimedia, the Internet, and printers. This is a useful introduction to how computers work that contains easy-to-understand text and gives the background computers users need to understand their computers. Large colorful illustrations, photographs, and cutaway diagrams help explain the text. An index and a computer CD are included.

Wright, Michael, and Mukul Patel, eds. *Scientific American: How Things Work Today*. New York: Crown Publishers, 2000. 288 pp. (hc. 0-375-41023-6) $29.95.
Gr. 9–12. Photographs, three-dimensional diagrams, exploded views, and succinct narrative explain how things work. Sections in the book are urban and domestic environment; communications and leisure; transportation; crime and security; power and industry; medicine and research; and space. References, basic scientific principles, a glossary, useful Internet addresses, and an index are included.

Chapter 16

The Future

The nonfiction books in this chapter examine robots, time travel, bionics, and artificial intelligence. Grounded in the present, these books look to the future and explore possibilities. They contain glimpses of what life will be like in the near and distant future.

Fowler, Allan. *It Could Still Be a Robot.* New York: Children's Press, 1997. 31 pp. (hc. 0-516-20431-9) $19.00; (pbk. 0-516-26258-0) $4.95.
Gr. P–2. Large-print text and colorful photographs introduce young readers to the world of robots in an easy to read format. A glossary and an index are included.

Beechcroft, Simon. *Super Humans: A Beginner's Guide to Bionics.* Illus. by Stephen Sweet and Stuart Squires. Brookfield, Conn.: Copper Beech Books, 1998. 32 pp. (hc. 0-7613-0621-8) $23.40.
Gr. 1–5. Color photographs and illustrations fill the pages of this exciting peek into the future as humans are transformed into super humans. Cloning, freezing bodies, and digital brainpower are some of the ideas examined in the book. "Reality Check" boxes are included for each life-prolonging method. These boxes contain information about the feasibility, the soundness of the science, the affordability, and how soon these new methods might be available. The book concludes with a glossary and an index.

Darling, David. *Could You Ever Build a Time Machine?* Minneapolis: Dillon, 1991. 60 pp. (hc. 0-87518-456-1) $14.95.

Gr. 3–6. This intriguing question starts students' minds to wondering. A brief discussion of time is followed by a discussion of time travel, Einstein's theory of relativity, and information on black holes and wormholes. Photographs, drawings, charts, diagrams, maps, instructions for creating a time capsule, a glossary, and an index are included.

Angliss, Sarah. *Cosmic Journeys: A Beginner's Guide to Living in Space.* Illus. by Stephen Sweet. Brookfield, Conn.: Copper Beech Books, 1998. 32 pp. (hc. 0-7613-0822-9) $23.90; (pbk. 0-7613-0741-9) $8.95.
Gr. 4–6. From the Future Files series. Photographs, drawings, movie stills, and brief chunks of text offer a captivating look at what it would take to build and maintain a city in space. Readers can make comparisons between where they live and a future city built in space. A glossary and an index are located at the end of the book.

Angliss, Sarah, and Colin Uttley. *Cities in the Sky: A Beginner's Guide to Living in Space.* Brookfield, Conn.: Copper Beech Books, 1998. 32 pp. (hc. 0-7613-0602-X) $23.90; (pbk. 0-7613-0635-8) $8.95.
Gr. 4–6. From the Future Files series. Photographs, drawings, movie stills, and abbreviated text describe what it might be like to live in space for extended periods. Albert Einstein and Stephen Hawking and their theories are briefly mentioned in the book. A glossary and an index are located at the end of the book.

Hellman, Hal. *Beyond Your Senses: The New World of Sensors.* New York: Dutton, 1997. 87 pp. (hc. 0-525-67533-7) $16.99.
Gr. 4–6. Intelligent machines need sensory capabilities similar to those of humans and this book discusses research on sensors that mimic humans' senses of touch, vision, smell, hearing, and taste. One chapter in the book focuses on the body's other sensors, such as internal ones that tell us when we have had enough to eat. A glossary, resources for learning more, and an index are included.

Bergin, Mark. *Robots.* New York: Franklin Watts, 2001. 32 pp. (hc. 0-531-14616-2) $27.50; (pbk. 0-531-14808-4) $9.95.
Gr. 4–8. From the Fast Forward series. This brief introduction to the world of robots has pages filled with colorful illustrations and short paragraphs of text. Bergin describes how robots work in hazardous situations, in the film industry, as robo-docs, and in space. A glossary, robot facts, and an index conclude the book.

Gifford, Clive. *How to Build a Robot*. Illus. by Tim Benton. Danbury,
     Conn.: Franklin Watts, 2001. 95 pp. (hc. 0-531-14649-9) $14.50;
     (pbk. 0-531-13997-2) $4.95.
Gr. 4–8. From the How To series. Readers learn what makes robots
work as they complete a series of simple experiments. Along the way,
they read about the history and the future of robots as well as compare
themselves to robots. Humorous black and white line drawings accompany the text.

Jefferis, David. *Artificial Intelligence: Robotics and Machine Evolu-
     tion*. New York: Crabtree, 1999. 32 pp. (hc. 0-7787-0046-1)
     $22.60; (pbk. 0-7787-0056-9) $8.95.
Gr. 4–8. From the Megatech series. This book contains a visual overview of the history of artificial intelligence and robotics including a
look into the future. A timeline, a glossary, and an index conclude the
book.

Krull, Kathleen. *They Saw the Future: Oracles, Psychics, Scientists,
     Great Thinkers, and Pretty Good Guessers*. Illus. by Kyrsten
     Brooker. New York: Atheneum Books for Young Readers, 1999.
     108 pp. (hc. 0-689-81295-7) $19.99.
Gr. 4–8. The predictions of the Oracle at Delphi, the Sibyls, the Maya,
Hildegard of Bingen, Leonardo da Vinci, Nostradamus, Jules Verne,
Nicholas Black Elk, H. G. Wells, Edgar Cayce, Jeane Dixon, and Marshall McLuhan are all described in this informative book that leaves it
to readers to make their own judgments about the predictions. Books
for further reading and index conclude the book.

Sonenklar, Carol. *Robots Rising*. Illus. by John Kaufmann. New York:
     Henry Holt, 1999. 103 pp. (hc. 0-8050-6096-0) $15.95.
Gr. 4–8. Robots perform hazardous work, go places humans cannot,
come in a variety of shapes and sizes, and these are just a few of the
things readers learn from this compact book. There is enough information here to get students excited and engaged in a study of what robots
can do and will do in the future. The book contains illustrations, photographs, a list of websites with information on building robots, a glossary, a bibliography, and an index.

Tambini, Michael. *Future*. New York: Alfred A. Knopf, 1998. 60 pp.
     (hc. 0-679-89317-2) $19.00.

Gr. 4–8. The book begins with timelines of inventions and events in the 1940s, the 1950s, the 1960s, the 1970s, the 1980s, and the 1990s. From the environment to the home to the human body, the book presents a wealth of information about the technology available today and what the future holds. Juxtaposed across two-page spreads are brief text blurbs and color photographs.

Baker, Christopher W. *Robots among Us: The Challenges and Promises of Robotics*. Brookfield, Conn.: The Millbrook Press, 2002. 48 pp. (hc. 0-7613-1969-7) $23.90.
Gr. 5–8. Baker introduces robots by asking readers if they would like a personal assistant to do the things they do not want to do such as taking out the garbage and mowing the lawn. Filled with questions and examples this book has students actively engaged as they explore robots. The book has a list of Internet resources and an index.

Gifford, Clive. *How the Future Began: Everyday*. New York: Kingfisher, 2000. 63 pp. (hc. 0-7534-5268-5) $15.95.
Gr. 5–8. From the How the Future Began series. The four major sections of the book cover these topics: cities and homes; transportation; work and play; and health issues. This exciting book is filled with photographs, illustrations, and just enough text to entice readers to become curious about what the future holds. A glossary, resources for learning more, and an index can be found at the end of the book.

——. *How the Future Began: Machines*. New York: Kingfisher, 1999. 63 pp. (hc. 0-7534-5188-3) $15.95.
Gr. 5–8. From the How the Future Began series. Peer into the future and see what is in store for industrial machines, military machines, domestic machines, machines in space, and the power needed to operate these machines. "Blurred Vision" textboxes describe past predictions that did not happen. "Crystal Ball" textboxes contain predictions for the distant future such as 2040 and 2100. A glossary, resources for learning more, and an index can be found at the end of the book.

Davies, Paul. *How to Build a Time Machine*. New York: Penguin Putnam, 2002. 132 pp. (hc. 0-670-03063-5) $19.95.
Gr. 6–12. Explore the possibility of constructing a traversable wormhole in order to create a time machine. The ideas about time travel presented in this book are sure to get readers thinking about the possibili-

ties. The book begins with a brief timeline of the history of time travel and concludes with an index.

Gott, J. Richard. *Time Travel in Einstein's Universe: The Physical Possibilities of Travel through Time*. Boston: Houghton Mifflin, 2001. 291 pp. (hc. 0-395-95563-7) $25.00.
Gr. 6–12. H. G. Well's *The Time Machine*, the movie *Back to the Future*, the works of Albert Einstein, George Gamow, Carl Sagan, and Stephen Hawking are all examined in this intriguing mix of science fiction and science. This is a book to set minds to wondering and is sure to engender lively classroom discussions. Black and white illustrations, notes, annotated references, and an index are included.

Wickelgren, Ingrid. *Ramblin' Roots: Building a Breed of Mechanical Beasts*. New York: Franklin Watts, 1997. 143 pp. (hc. 0-531-11301-9) $23.00; (pbk. 0-531-15829-2) $9.95.
Gr. 6–12. Beginning with a history of robots, Wickelgren then describes current robots and the jobs they perform. She then tells how scientists are working to create robots that can mimic human and animal behavior. Robot resources, a glossary, source notes, a bibliography, and an index are included.

Menzel, Peter, and Faith D'Aluisio. *Robo Sapiens: Evolution of a New Species*. Cambridge, Mass.: The MIT Press, 2000. 239 pp. (hc. 0-262-13382-2) $29.95.
Gr. 9–12. Look into the future of intelligent robots and ponder whether or not they will be our equals. Pioneers in the field of robotics take readers inside their world to contemplate the future of robots. Large color photographs, a glossary, and an index are included.

Stix, Gary, and Miriam Lacob. *Who Gives a Gigabyte?: A Survival Guide for the Technologically Challenged*. New York: John Wiley and Sons, 1999. 300 pp. (hc. 0-471-16293-0) $24.95.
Gr. 9–12. The book provides clear, concise descriptions of a variety of technological advances in areas including computers, lasers, genes, and the environment. While the authors readily acknowledge they do not cover everything, they cover enough advances to satisfy those curious about many present and future technologies. Books for further reading and an index are included.

Cetron, Marvin, and Owen Davies. *Probable Tomorrows: How Science and Technology Will Transform Our Lives In the Next Twenty Years.* New York: St. Martin's Press, 1997. 298 pp. (hc. 0-312-15429-1) $24.95.
Gr. 10–12. Personal computers, transportation, consumer goods, heavy industries, and medical advances are some of the areas covered in this venture into the future. The book concludes with a timetable for the future and an index.

Gershenfeld, Neil. *When Things Start to Think.* New York: Henry Holt, 1999. 225 pp. (hc. 0-8050-5874-5) $25.00.
Gr. 10–12. Step into the future and behold the possibilities technology has to offer as described by Gershenfeld. From shoes that exchange data through a handshake to books that reconfigure themselves, these are examples of future technology that will be responsive to individual user's needs. An afterword and an index conclude the book.

Moravec, Hans. *Robot: Mere Machine to Transcendent Mind.* New York: Oxford University Press, 1999. 227 pp. (hc. 0-19-511630-5) $25.00.
Gr. 10–12. Moravec explores the world of artificial intelligence and robots. This provocative book predicts that by the year 2050 the intelligence of robots will exceed that of humans. This engaging narrative on an absorbing topic causes readers to stop and ponder the future. Notes and an index conclude the book.

Nolte, David D. *Mind at Light Speed: A New Kind of Intelligence.* New York: The Free Press, 2001. 303 pp. (hc. 0-7432-0501-4) $26.00.
Gr. 10–12. Rather than computers using electricity, imagine computing with light. These optical computers would operate at the speed of light, be extremely efficient, and generate a new form of intelligence. An appendix lists the optical research and development companies that are supporting this optical revolution. Appendixes, a glossary, a bibliography, notes, and an index are provided.

Hazen, Robert, with Maxine Singer. *Why Aren't Black Holes Black?: The Unanswered Questions at the Frontiers of Science.* New York: Doubleday, 1997. 309 pp. (pbk. 0-385-48014-8) $12.95.
Gr. 11–12. Questions in the book are from the scientific fields of earth science, chemistry, physical science, and biochemistry. The authors discuss past research and ongoing research that seeks to find answers to

the questions. For readers interested in learning more about the questions discussed in the book there are lists of resources for each question at the end of the book.

Maddox, John. *What Remains to Be Discovered?: Mapping the Secrets of the Universe, the Origins of Life and the Future of the Human Race*. New York: Simon & Schuster, 1998. 434 pp. (hc. 0-684-82292-X) $26.00; (pbk. 0-684-86300-6) $15.00.
Gr. 11–12. Maddox writes about the areas of science likely to generate the most crucial discoveries in the future. He organizes his ideas into three areas: matter, life, and our world. He ends by reminding readers that in some areas of inquiry progress is noted not by the discoveries but by the generation of more complex questions. Notes and an index are provided.

Matathia, Ira, and Marian Salzman. *Next: Trends for the Near Future*. Woodstock, N.Y.: The Overlook Press, 1999. 414 pp. (hc. 0-87951-943-6) $26.95.
Gr. 11–12. This is an international look at future trends predicted for the next five to ten years. Since the book was published in 1999, students can now read the book and determine if the predictions were accurate. When the book was published, the millennium bug was pending and cause for great concern. For the most part it caused few problems, but it is mentioned frequently in the book as a potential problem for the future.

Chapter 17

Reference

The selected reference sources in this chapter encompass topics such as science, weapons, everyday life, medicine, transportation, how things are made, and how things work. Dictionaries, encyclopedias, timelines, and yearbooks are included in these annotations.

Rubel, David. *Science*. New York: Scholastic, 1995. 192 pp. (hc. 0-590-49367-1) $18.95.
Gr. 2–4. From the Kid's Encyclopedia series. This encyclopedia brims with photographs, diagrams, illustrations, and sidebars—all enhancing the concise text. The categories include astronomy; biology; earth science; the human body; and physics and chemistry. This one-volume encyclopedia has a glossary and an index.

Berger, Melvin. *Scholastic Science Dictionary*. Illus. by Hannah Bonner. New York: Scholastic, 2000. 224 pp. (hc. 0-590-31321-5) $19.95.
Gr. 3–6. Included with more than two thousand four hundred entries are very brief biographies of key scientists. Terms defined in the book come from astronomy, biology, chemistry, geology, paleontology, and physics. Colorful illustrations, diagrams, and an index are included.

Hall, Roland, ed. *Inside a. . .* 16 vols. Danbury, Conn.: Grolier Educational, 2001. 512 pp. (hc. 0-7172-9521-4) $299.00.
Gr. 4–8. The volumes in this set include car, clock, compact disk, computer, construction machine, helicopter, high-speed train, jet plane, powerboat, rocket, satellite, skyscraper, stove, telephone, television, and website. Photographs, diagrams, and "fact files" help readers un-

derstand the inner workings of these inventions. Each volume has its own glossary, resources for learning more, and an index.

Rose, Sharon, and Neil Schlager. *CDs, Super Glue, and Salsa: How Everyday Products Are Made.* 2 vols. New York: Gale, 1995. 288 pp. (hc. 0-8103-9791-9) $99.00.
Gr. 4–8. From the automobile to the zipper, these alphabetically arranged entries describe how thirty everyday objects are made. Black and white photographs and labeled diagrams aid in understanding the explanations in the text. A set index can be found in both volumes.

Bruno, Leonard C. *Science and Technology Firsts.* Detroit: Gale, 1997. 636 pp. (hc. 0-7876-0256-6) $90.00.
Gr. 4–12. Chapters in the book include agriculture and everyday life; astronomy; biology; chemistry; communications; computers; earth science; energy, power systems, and weaponry; mathematics; medicine; physics; and transportation. Within each chapter the entries are listed chronologically. A bibliography and an index conclude the book.

Daintith, John, and Derek Gjertsen, eds. *The Grolier Library of Science Biographies.* 10 vols. Danbury, Conn.: Grolier Educational, 1997. 2,530 pp. (hc. 0-7172-7626-0) $329.00.
Gr. 4–12. The entries contain basic biographic data, but the main focus is on the individuals' achievements and the importance of these achievements. Small black and white photographs accompany many of the entries. Each volume contains a list of resources for further reading, a glossary, and an index.

Schmittroth, Linda, Mary Reilly McCall, and Bridget Travers, eds. *Eureka! Scientific Discoveries and Inventions that Shaped the World.* 6 vols. New York: Gale, 1995. 1,179 pp. (hc. 0-8103-9802-8) $165.00.
Gr. 4–12. These six volumes include six hundred entries pertaining to scientific inventions and discoveries that have changed the world. Each entry notes the person or persons responsible for the discovery or invention, gives information on the knowledge and technology used, and explains its impact. The entries are grouped into thirty-seven categories and within the categories they are arranged alphabetically. A master index is included in each volume.

Yount, Lisa. *A to Z of Women in Science and Math*. New York: Facts
On File, 1999. 254 pp. (hc. 0-8160-3797-3) $44.00.
Gr. 4–12. An author's note at the beginning of the book states that
women were chosen for the volume based on their contributions, fame,
and diversity. The introduction discusses the obstacles overcome by
women scientists and mathematicians. Entries describe the women's
contributions and include resources for further reading. Sources, entries
indexes, a chronology, and a general index are provided.

Adams, Simon. *Visual Timeline of the Twentieth Century*. New York:
DK Publishing, 1996. 48 pp. (hc. 0-7894-0997-6) $15.95.
Gr. 5–9. The events in each chapter are organized into four categories
that span the pages. The categories are arts and entertainment; science
and discovery; everyday life; and world events. The layout of the book
includes photographs and reproductions with short text entries written
in small print. An index concludes the book.

Platt, Richard. *Smithsonian Visual Timeline of Inventions*. New York:
DK Publishing, 1994. 64 pp. (hc. 1-56458-675-8) $15.95.
Gr. 5–9. Over four hundred inventions are included on this visual timeline. As the timeline weaves across the pages, a chronology of world
events unfolds along the bottom of the pages, which helps the reader
place the inventions in the context of the times. The book concludes
with an index of inventions and an index of inventors.

Knight, Judson, and Neil Schlager. *Science of Everyday Things*. 4 vols.
New York: Gale, 2002. 1,600 pp. (hc. 0-7876-5631-3) $299.00.
Gr. 5–10. The set includes a volume on each of these four topics:
chemistry, physics, biology, and earth science. Brief entries provide
basic information and some illustrations are included. The set includes
a subject index, a cumulative index of "everyday things," and a cumulative general subject index.

Rubel, David. *Scholastic Timelines: The United States in the Nineteenth Century*. New York: Scholastic, 1995. 192 pp. (hc. 0-590-
72564-5) $16.95.
Gr. 5–10. The chapters are organized by eras, and within each chapter
the events are placed into one of four categories: politics; life in the
nineteenth century; arts and entertainment; and science and technology.
The right-hand pages highlight in pictures and words an event or person

who was important to that era. The book includes photographs, illustrations, a glossary, and an index.

——. *Scholastic Timelines: The United States in the Twentieth Century.*
New York: Scholastic, 1995. 192 pp. (hc. 0-590-27134-2) $16.95.
Gr. 5–10. The chapters are organized by eras, and within each chapter the events are placed into one of four categories: politics; life in the twentieth century; arts and entertainment; and science and technology. On the right side of the two-page spread are feature articles about people or events that helped to shape the era. This is a useful reference source for students as they begin research projects. The book includes photographs, illustrations, a glossary, and an index.

The New Book of Popular Science. 6 vols. New York: Grolier, 2002.
3,798 pp. (hc. 0-7172-1223-8) $269.00.
Gr. 5–12. This book contains current information these subject areas: astronomy, space science, mathematics, earth science, energy, environmental sciences, chemistry, physics, biology, plant life, animal life, mammals, human sciences, and technology. Color photographs, diagrams, maps, and sidebars with science activities accompany the concise entries. Included in each volume is a full set index.

Popular Science: Science Year by Year—Discoveries and Inventions from the Twentieth Century that Shape Our Lives Today. New York: Scholastic, 2001. 240 pp. (hc. 0-439-284380-4) $19.95.
Gr. 5–12. Students looking for a quick reference source for discoveries and inventions find this book helpful. Photographs, diagrams, sidebars, science news boxes, and brief descriptive text explain the wondrous innovations. The book contains an extensive index.

Bunch, Bryan H., and Alexander Hellemans. *The Timetables of Technology: A Chronology of the Most Important People and Events in the History of Technology.* New York: Simon & Schuster, 1993. 490 pp. (hc. 0-671-76918-9) $35.00.
Gr. 6–12. Grouped according to historical eras, each grouping provides a brief overview of the time period and a description of major advances. Essays and biological profiles supplement the entries. Name and subject indexes conclude the book.

Cohen, I. Bernard. *From Leonardo to Lavoisier, 1450–1800.* New York: Charles Scribner's Sons, 1998. 303 pp. (hc. 0-684-15377-7) $90.00.
Gr. 6–12. From the Album of Science series. The focus of this volume is on the establishment of modern science. Astronomy, physical science, life science, and earth science are represented. Reproductions and technical drawings highlight the text. A bibliography and a guide to further reading as well as an index conclude the book.

Darling, David. *The Complete Book of Spaceflight From Apollo One to Zero Gravity.* Hoboken, N.J.: John Wiley and Sons, 2003. 537 pp. (hc. 0-471-05649-9) $35.00.
Gr. 6–12. More than three thousand entries describe spaceflight technology, people, and events. From the ancient Greeks to traveling at warp speed, Darling covers the past, the present, and the future in this reference book. Cross-references, charts, photographs, and sidebars contain supplemental material. A list of acronyms and abbreviations, references, websites, and a category index are provided.

Day, Lance, and Ian McNeil, eds. *Biographical Dictionary of the History of Technology.* New York: Routledge, 1996. 844 pp. (hc. 0-415-06042-7) $195.00.
Gr. 6–12. Almost one thousand three hundred individuals are recognized in this dictionary for their significant contributions to the advancement of technology. Many of the entries end with suggested resources for obtaining more information. Entries are indexed by subject area, topics, and names.

de Bono, Edward. *Eureka! An Illustrated History of Inventions from the Wheel to the Computer.* New York: Holt, Rinehart and Winston, 1974. 248 pp. (hc. 0-03-012641-X) $25.00.
Gr. 6–12. Entries are arranged under these broad headings: man moving, man taking, man living, man working, and key devices. Under each heading is an introduction that provides a broad overview of the topic. There are subheadings under each of the headings that aid in organizing the entries. Photographs and illustrations accompany the entries. A chronological table, an index of names, and an index of subjects complete the book.

History of Science. 10 vols. New York: Grolier, 2003. 800 pp. (hc. 0-7172-5729-0) $279.00.

Gr. 6–12. This encyclopedia explores the development of science and introduces readers to scientists, mathematicians, alchemists, and philosophers as well as their explorations and experiments. The ten volumes are science in ancient civilizations; Islamic and Western medieval science; traditions of science outside of Europe; the European Renaissance; the scientific revolution; the eighteenth century; physical science in the nineteenth century; biology and geology in the nineteenth century; atoms and galaxies; modern physical science; and twentieth-century life science. Some of the entries include photographs, diagrams, maps, charts, timelines, and sidebars. Each volume contains resources for learning more, a glossary, a chronology, and a set index.

How It Works: Science and Technology, 3rd ed. 10 vols. Tarrytown, N.Y.: Marshall Cavendish, 2003. 2,880 pp. (hc. 0-7614-7314-9) $499.95.

Gr. 6–12. Clear explanations, cutaway diagrams, and colorful illustrations assure that students understand the discoveries and inventions discussed in these reference volumes. Many entries include "See Also" text boxes, "Fact File" boxes, and cross-references. Volume twenty contains a general index, subject indexes, and a glossary.

L'E Turner, Gerard. *Scientific Instruments 1500–1900: An Introduction.* Berkeley, Calif.: University of California Press, 1998. 144 pp. (hc. 0-520-21728-4) $40.00.

Gr. 6–12. Studying the scientific instruments of the past four centuries provides information on the development of science during this time period. Astronomy, navigation, optical, surveying, and medical instruments are some of the types included in the book. Photographs and reproductions add to the text and many of them are full color. A bibliography, a listing of museums and collections, and an index are provided.

McGrath, Kimberley A., ed. *World of Invention,* 2nd ed. Detroit: Gale, 1999. 1,043 pp. (hc. 0-7876-2759-3) $115.00.

Gr. 6–12. This book focuses on inventions beginning with the start of the Industrial Revolution to the present. Some entries reflect the evolution of the invention over time such as the entry for aircraft. A bibliography, a subject index, and a general index are contained in this book.

——. *World of Scientific Discovery,* 2nd ed. Detroit: Gale, 1999. 1,186
   pp. (hc. 0-7876-2760-7) $115.00.
Gr. 6–12. Over one thousand three hundred entries on scientific milestones and scientists from the late sixteenth century to the present are
contained in this one volume. Entries include background information
on the scientific discoveries, an explanation of the discovery, and information on its scientific impact. A bibliography, a subject index, and
a general index are provided.

Mount, Ellis, and Barbara A. List. *Milestones in Science and Technology: The Ready Reference Guide to Discoveries, Inventions, and
   Facts,* 2nd ed. Phoenix, Ariz.: Oryx, 1994. 206 pp. (hc. 0-89774-
   671-6) $39.50.
Gr. 6–12. Over one thousand two hundred fifty entries about important
events and inventions are contained in this slim volume. The short concise entries include cross-references. An alphabetical listing of terms, a
listing of references, and several indexes are included.

Saari, Peggy, and Stephen Allison, eds. *Scientists: The Lives and Works
   of One Hundred Fifty Scientists.* 3 vols. New York: Gale, 1996.
   1,045 pp. (hc. 0-7876-0959-5) $145.50.
Gr. 6–12. From the Industrial Revolution to the present, these are the
scientists whose major breakthroughs created an enduring legacy. The
entries include subheadings, "impact" boxes that explain why the scientists' work is important, biographical boxes that describe scientists
who conducted similar work, and a list of books for further reading. A
timeline, a glossary, and an index are included.

Smith, Roger, ed. *Inventions and Inventors.* Hackensack, N.J.: Salem
   Press, 2002. 936 pp. (hc. 1-58765-016-9) $104.00.
Gr. 6–12. This two-volume collection of essays describes one hundred
ninety-five noteworthy inventions from the twentieth century. The essays range in length from three to five pages and encompass diverse
topics. Biographical sidebars, sources for further study, cross-
references, a timeline, a category list of entries, and a subject index are
included.

Travers, Bridget, and Fran Locher Freiman, eds. *Medical Discoveries:
   Medical Breakthroughs and the People Who Developed Them.* 3
   vols. Detroit: Gale, 1997. 429 pp. (hc. 0-7876-0890-4) $125.00.

Gr. 6–12. Medical and dental discoveries and inventions fill this well-written resource. The entries include information on the individuals responsible for the breakthrough and the impact of the breakthrough on the field of medicine. Sidebars, "See-Also" references, photographs, illustrations, a timeline of medical events, a glossary, a bibliography, and a master index are included.

Web, Pauline, and Mark Suggitt. *Gadgets and Necessities: An Encyclopedia of Household Innovations*. Santa Barbara, Calif.: ABC–CLIO, 2000. 377 pp. (hc. 1-57607-081-6) $75.00.
Gr. 6–12. Focusing on twentieth-century innovations, this is a study of the material culture that was shaped by society and in turn shaped society. Black and white photographs, a list of designers, an alphabetical listing of innovations, a technical glossary, and an index are provided.

Webster, Raymond B. *African American Firsts in Science and Technology*. Detroit: Gale, 1999. 462 pp. (hc. 0-7876-3876-5) $75.00.
Gr. 6–12. These concise entries highlight the accomplishments ranging from noted inventor Benjamin Banneker to Ellwood G. Ivey, inventor of the bio-system sensitized steering wheel. This steering wheel detects if the driver is intoxicated and disables the vehicle when the driver's blood alcohol exceeds the legal limit. The book contains subject chapters with entries arranged chronologically from 1706 to the present. Included in the book are a bibliography, an index by year, an occupation index, and a general index.

Weiner, Richard. *Webster's New World Dictionary of Media and Communications*. New York: Macmillan, 1996. 676 pp. (pbk. 0-02-860611-6) $39.95.
Gr. 6–12. Although this is a specialized dictionary, the entries are clear and easy-to-understand. The over thirty-five thousand entries range from slang to technical terms. Cross-references are included.

World Book Biographical Encyclopedia of Scientists. 8 vols. Chicago: World Book, 2002. 1,536 pp. (hc. 0-7166-7600-1) $289.00.
Gr. 6–12. These biographies of over one thousand three hundred scientists from around the world include photographs, illustrations, and cross-references. Also included are feature articles covering major scientific discoveries and achievements. Volume eight contains a glossary, several indexes, a list of Nobel Prize winners, and other resources.

Zierdt-Warshaw, Linda, Alan Winkler, and Leonard Bernstein. *American Women in Technology: An Encyclopedia.* Santa Barbara, Calif.: ABC–CLIO, 2000. 384 pp. (hc. 0-57607-072-7) $75.00.
Gr. 6–12. The encyclopedia begins with a discussion of the traditional and societal barriers that women have faced throughout history. The technological contributions of women include architecture, engineering, mathematics, and other scientific disciplines. Entries include brief biographical information, a description of the women's contributions, "See-Also" entries, and references. Some of the entries include photographs. An appendix of award winners, a bibliography, and an index are provided.

Brown, Travis. *Historical First Patents: The First United States Patent for Many Everyday Things.* Metuchen, N.J.: Scarecrow Press, 1994. 216 pp. (hc. 0-8108-2898-7) $41.50.
Gr. 8–12. The introduction to the book incorporates information on United States patent laws and the patent office. Each entry contains a short history of the invention, brief biographical information on the inventor, and the process involved in creating the invention. Entries are arranged alphabetically and include sketches or diagrams. A bibliography and an index are included.

Culligan, Judy, ed. *Scientists and Inventors.* New York: Macmillan Library Reference, 1999. 389 pp. (hc. 0-02-864983-4) $90.00.
Gr. 8–12. From the Macmillan Profiles series. Concise profiles of over one hundred scientists and inventors are presented in this volume. Brief timelines of the scientists' lives and definitions of key words in the text are found in the margins of the entries. A glossary and an index conclude the book.

The Cutting Edge: An Encyclopedia of Advanced Technologies. New York: Oxford University Press, 2000. 360 pp. (hc. 0-19-512899-0) $75.00.
Gr. 8–12. Entries contain a brief introduction, subheadings including scientific and technological descriptions; historical developments; uses, effects, and limitations; and issues and debates. Entries conclude with related topics, a bibliography, and resources for further research. Photographs, diagrams, and cutaway views are included for some of the entries. An index is provided.

Olson, James Stuart. *Encyclopedia of the Industrial Revolution.* Westport, Conn.: Greenwood Press, 2002. 313 pp. (hc. 0-313-30830-6) $69.95.

Gr. 8–12. The book focuses on the 1750s to the 1920s. Entries cover a wide range of topics including among others technologies, inventions, natural resources, and court cases. Cross-references are indicated in bold text and many entries include suggestions for further reading. A chronology of the Industrial Revolution in America, a bibliography, and an index are provided.

Pfaffenberger, Bryan. *Webster's New World Dictionary of Computer Terms,* 8th ed. Foster City, Calif.: International Data Group (IDG) Books Worldwide, 2000. 588 pp. (pbk. 0-02-863777-1) $12.95.

Gr. 8–12. Definitions cover both computer and Internet terminology. Many of the entries contain additional background information beyond the definitions.

Ralston, Anthony, Edwin D. Reilly, and David Hemmendinger. *Encyclopedia of Computer Science,* 4th ed. New York: Grove Dictionaries, 2000. 2,034 pp. (hc. 1-561-59248-X) $50.00.

Gr. 8–12. In addition to the encyclopedia entries, there are articles, biographies, diagrams, and photographs. Some entries include cross-references and a bibliography. Articles are grouped under these headings: hardware, software, computer systems, information and data, mathematics of computing, theory of computation, methodologies, applications, and computing milieu. Several appendixes, a name index, and a subject index are provided.

Travers, Bridget, ed. *The Gale Encyclopedia of Science.* 6 vols. New York: Gale Research, 1996. 4136 pp. (hc. 0-8103-9841-9) $526.50.

Gr. 8–12. Entries in this encyclopedia range from short definitions to longer explorations that contain photographs and diagrams. Sidebars contain the definitions of key terms found in the entries. "See-Also" references and cross-references provide additional resources for readers. A general index is included.

Trefil, James, ed. *Encyclopedia of Science and Technology.* New York: Routledge, 2001. 554 pp. (hc. 0-415-93724-8) $50.00.

Gr. 8–12. These succinct entries cover key aspects of science, medicine, and technology. This encyclopedia has a "Critical Path" feature that enables students to locate information on an entire discipline. Sidebars and illustrations are found throughout the volume. Bibliographic references and an index conclude the book.

Women in Science. New York: Macmillan Library Reference, 2001.
  421 pp. (hc. 0-02-865502-8) $90.00.
Gr. 8–12. From the Macmillan Profiles series. are These concise biographies and some include photographs. Brief timelines of the scientist's life and definitions of key words in the text are found in the margins of the entries. A timeline, sources, additional resources, a glossary, and an index conclude the book.

Anzovin, Steven, and Janet Podell. *Famous First Facts: Thousands of First Happenings, Discoveries and Inventions That Have Occurred throughout the World.* New York: H.W. Wilson, 2000. 837 pp. (hc. 0-8242-0958-3) $150.00.
Gr. 9–12. The over five thousand entries contain "firsts" from hundreds of countries ranging from 3.5 billion years ago to 2001. Entries are arranged alphabetically under broad categories and subcategories. Each entry has a number to assist readers when using the five indexes. The indexes included in this book are subject matter, year, day, personal name, and geographical location.

Bull, Stephen. *An Historical Guide to Arms and Armor.* New York: Facts On File, 1991. 224 pp. (hc. 0-8160-2620-3) $35.00.
Gr. 9–12. These five chronological chapters include arms and armor ranging from the Greeks and Romans to the nineteenth century. Over three hundred illustrations, an index, and a bibliography are included.

Emsley, John. *Nature's Building Blocks: An A–Z Guide to the Elements.* New York: Oxford University Press, 2001. 539 pp. (hc. 0-19-850341-5) $29.95.
Gr. 9–12. This alphabetically arranged guide includes all one hundred fifteen elements. Entries include information such as when and how the element was discovered; where it comes from; and its uses and misuses. An appendix of the discovery of the elements in chronological order, a bibliography, a list of elements and atomic numbers, and a periodic table are provided.

Finniston, Monty, ed. *Oxford Illustrated Encyclopedia of Invention and Technology.* New York: Oxford University Press, 1992. 391 pp. (hc. 0-19-869138-6) $49.95.

Gr. 9–12. This volume encompasses a broad overview of modern technology including alternative technologies and renewable energy
sources. The alphabetically arranged entries include cross-references,
diagrams, charts, and photographs.

Harding, David, ed. *Weapons: An International Encyclopedia from
5000 B.C. to 2000 A.D.* New York: St. Martin's Press, 1990. 336
pp. (hc. 0-312-03951-4) $27.95.
Gr. 9–12. Photographs, illustrations, and diagrams accompany the entries that chronicle the evolution of weapons and explain how they
work. The weapons are grouped by categories, including hand weapons; hand-thrown missiles; hand-held missile-throwers; mounted missile-throwers; positioned weapons; bomb and self-propelled missiles;
chemical, nuclear, and biological weapons; and the thinking battlefield.
A bibliography and an index are provided.

Pagel, Mark, ed. *Encyclopedia of Evolution.* 2 vols. New York: Oxford
University Press, 2002. 1,205 pp. (hc. 0-19-512200-3) $325.00.
Gr. 9–12. These two volumes contain alphabetical entries and overview
essays on evolution. Sidebars and diagrams are interspersed throughout
the text and a topical outline of articles can be found at the beginning of
the book. An index is provided.

Paine, Lincoln P., with James H. Terry and Hal Fessenden. *Ships of the
World: An Historical Encyclopedia.* Boston: Houghton Mifflin,
1997. 680 pp. (hc. 0-395-71556-3) $50.00.
Gr. 9–12. More than one thousand important or well-known vessels are
described in this encyclopedia. Entries begin with statistics on the vessel followed by a description of the vessel and why it was either important or well known. The encyclopedia includes maps, a listing of
ships in literature, chronologies, a glossary, a bibliography, and an index.

Simmons, John. *The Scientific One Hundred: A Ranking of the Most
Influential Scientists.* Secaucus, N.J.: Carol Publishing Group,
1996. 504 pp. (hc. 0-8065-1749-2) $29.95.
Gr. 9–12. The scientists profiled in this book are noted for their discoveries of new things in nature; hence, many inventors are not included in
this volume. These are the top five scientists according to Simmon's
rankings: Isaac Newton, Albert Einstein, Niels Bohr, Charles Darwin,
and Sigmund Freud. The book includes an appendix of inexcusable

omissions, honorable mentions, and also-rans. Source notes, a bibliography, and an index conclude the book.

Stone, George Cameron. *A Glossary of the Construction, Decoration, and Use of Arms and Armor in All Countries and in All times: Together with Some Closely Related Subjects*. Mineola, N.Y.: Dover Publications, 1999. 693 pp. (pbk. 0-486-40726-8) $439.95.
Gr. 9–12. Stone's personal collection of arms and armor were used for much of the research needed for this book. The entries are organized alphabetically and many of them include captioned photographs. A bibliography is provided.

Weissman, Paul R., Lucy-Ann McFadden, and Torrence V. Johnson. *Encyclopedia of the Solar System*. San Diego: Academic, 1999. 992 pp. (hc. 0-12-226805-9) $99.95.
Gr. 9–12. This volume presents a comprehensive overview of the science of the solar system including the current state of knowledge and its evolution. The entries are arranged in forty chapters and include color plates, figures, and tables. An appendix and an index are included.

Williams, Trevor I. *A History of Invention: From Stone Age to Silicon Chips*. New York: Facts On File, 2000. 367 pp. (hc. 0-8160-4072-9) $45.00.
Gr. 9–12. From ancient civilizations to the present day, inventions have changed the way people live. This book shows how inventions in different parts of the world from different civilizations have influenced the technology we use today. Color photographs, diagrams, and drawings are included. The book concludes with a biographical dictionary, a bibliography, and an index.

Windelspecht, Michael. *Groundbreaking Scientific Experiments, Inventions, and Discoveries of the Seventeenth Century*. Westport, Conn.: Greenwood Press, 2002. 270 pp. (hc. 0-313-31501-9) $69.95.
Gr. 9–12. From the Groundbreaking Scientific Experiments, Inventions, and Discoveries through the Ages series. Windelspecht explains that in the seventeenth century scientists conducted experiments and wrote mathematical proofs whereas prior to this most scientific endeavors were based on observations. Entries are followed by selected bibliographies. A timeline, an appendix of entries by scientific field, a

glossary, a selected bibliography, a subject index, and a name index are
included.

Zimmerman, Robert. *The Chronological Encyclopedia of Discoveries
in Space.* Phoenix, Ariz.: Oryx, 2000. 410 pp. (hc. 1-57356-196-7)
$95.00.
Gr. 9–12. An introduction briefly describes discoveries before 1957.
The focus of the book is on discoveries after that time. The book provides detailed information on the experiments and investigations per-
formed during space missions. Diagrams and photographs are interspersed throughout the text. Appendixes, a glossary, a bibliography,
and an index conclude the book.

Bloomfield, Louis. A. *How Things Work: The Physics of Everyday
Life,* 2nd ed. New York: John Wiley and Sons, 2001. 512 pp. (pbk.
0-471-38151-9) $86.75.
Gr. 10–12. Written for nonscientists, this book makes connections between everyday experiences and science as it describes how things
work. Diagrams, illustrations, and photographs aid in understanding the
concepts. Appendixes, a glossary, and an index conclude the book.

King, Robert C., and William Stansfield. *A Dictionary of Genetics,* 6th
ed. New York: Oxford University Press, 2001. 544 pp. (hc. 0-19-
514324-8) $65.00; (pbk. 0-19-509442-5) $29.95.
Gr. 10–12. Entries include genetic and nongenetic terms often encountered when reading about genetics. Drawings and tables are included.
Several appendixes are found at the back of the book, including one
that contains a listing of discoveries and inventions related to genetics.

Levine, Harry III. *Genetic Engineering: A Reference Handbook.* Santa
Barbara, Calif.: ABC–CLIO, 1999. 264 pp. (hc. 0-87436-962-2)
$34.00.
Gr. 10–12. This reference book contains an overview of genetic engineering, a chronology, biographical sketches of those involved in genetics research, a directory of organizations, and listings of print and
nonprint resources. A glossary and an index conclude the book.

Moore, Sir Patrick, ed. *Astronomy Encyclopedia.* New York: Oxford
University Press, 2002. 456 pp. (hc. 0-19-531833-7) $50.00.
Gr. 10–12. Over one hundred astronomers contributed entries to this
comprehensive volume. The entries are arranged in alphabetical order

and some include cross-references. Photographs, star maps, sidebars, charts, and diagrams enhance the text.

Rosner, Lisa, ed. *Chronology of Science: From Stonehenge to the Human Genome Project.* Santa Barbara, Calif.: ABC–CLIO, 2002. 566 pp. (hc. 1-57607-954-6) $85.00.
Gr. 10–12. The chapters are organized into these categories: earliest discoveries, the Medieval World, the Scientific Revolution, the industrializing world, and the twentieth century. Within each chapter the entries are subdivided into smaller time periods and by subject area. Appendixes, a glossary, and an index are included.

Scientific American Desk Reference. New York: John Wiley and Sons, 1999. 690 pp. (hc. 0-471-35675-1) $45.00.
Gr. 10–12. The editors of *Scientific American* have compiled this concise overview of science and technology facts and anecdotes. Included are brief biographies, chronologies, cross-references, and discussions of key topics. An index concludes the book.

Chapter 18

Invention to Market

Books in this chapter provide inventors both young and old with information on how to design, produce, patent, and market their inventions. Patent attorneys wrote some of the books annotated below. Some of the attorneys advocate a do-it-yourself approach to applying for patents, copyrights, and trademarks; others strongly advocate hiring an attorney to complete the applications. Whichever route inventors choose, these books provide essential knowledge to help them understand the process. The Internet holds additional information and Appendix B Electronic Resources lists some helpful websites.

Caney, Steven. *Steven Caney's Invention Book.* New York: Workman Publishing, 1985. 207 pp. (pbk. 0-89480-076-0) $8.95.
Gr. 4–6. The book has two sections. The first section is an inventor's handbook that describes the process of inventing and the process of patenting ideas. The second part of the book contains great invention stories with ideas to get young inventors started on their own inventions. Black and white photographs, reproductions, sidebars, and diagrams and an index are included.

Erlbach, Arlene. *The Kids' Invention Book.* Minneapolis: Lerner Publications, 1997. 64 pp. (hc. 0-8225-2414-7) $22.60; (pbk. 0-8225-9844-2) $9.95.
Gr. 5–8. Readers meet eleven young inventors and their inventions and then learn about the invention process. The book contains information on invention contests, obtaining patents, and starting an inventor's club. Illustrations, resources for learning more, a glossary, and an index are included.

Coleman, Bob, and Deborah Neville. *The Great American Idea Book*. New York: W.W. Norton, 1993. 334 pp. (hc. 0-393-03447-X) $22.95.
Gr. 9–12. Turning great ideas into moneymaking ventures is the topic of this book. The authors explain how to protect, develop, promote, defend, and profit from ideas. Systematic plans, answers to common questions, and real-life examples make this a useful reference source for inventors. A recommended reading list and an index are included.

DeMatteis, Bob. *From Patent to Profit: Secrets and Strategies for the Successful Inventor*. Garden City Park, N.Y.: Avery Publishing, 1998. 288 pp. (pbk. 0-89529-879-1) $29.95.
Gr. 9–12. The book contains four sections filled with useful, practical advice for the inventor on inventing, patenting, manufacturing and marketing, and licensing. Diagrams, sample letters, anecdotes, and quotes complement the well-written text. An appendix of inventor resources, a strategic guide, and an index conclude the book.

Elias, Stephen, and Richard Stim. *Patent, Copyright, and Trademark*, 5th ed. Berkeley, Calif.: Nolo, 2002. 464 pp. (pbk. 0-87337-848-2) $39.99.
Gr. 9–12. The introduction explains how property law works and how to use the book. There are separate sections in the book for the four separate categories of intellectual property: trade secret law, copyright law, patent law, and trademark law. The book contains an overview of each category, short explanations of key terms, charts, lists, and copies of useful forms.

Gibbs, Andy, and Bob DeMatteis. *Essentials of Patents*. Hoboken, N.J.: John Wiley and Sons, 2003. 270 pp. (pbk. 0-471-25050-3) $29.95.
Gr. 9–12. Gibbs and DeMatteis describe patents as a "new monetary unit" and explain how that impacts inventors, corporations, and shareholders. An index of organizations and a subject index are included.

Gold, Robert J. *Eureka! The Entrepreneurial Guide to Developing, Protecting, and Profiting from Your Ideas*. Englewood Cliffs, N.J.: Prentice Hall, 1994. 281 pp. (pbk. 0-13-011735-8) $18.95.
Gr. 9–12. Inventor Gold shares his expertise in developing and marketing inventions in this useful guide. The book contains forms, checklists, and sample letters. Information on government agencies, inventor

associations, and university resources can be found in the appendix. The book includes a glossary and an index.

Kanbar, Maurice. *Secrets from an Inventor's Notebook.* San Francisco: Council Oak Books, 2001. 192 pp. (hc. 1-57178-099-8) $22.95.
Gr. 9–12. Writing from his own experiences, Kanbar explains his five steps to successful inventing. The steps include solving a problem, building a prototype, protecting the idea, deciding to manufacture or license, and marketing the idea. The appendix is filled with a variety of useful resources for inventors. A bibliography and an index conclude the book.

Mosely, Jr., Thomas E. *Marketing Your Invention,* 2nd ed. Chicago: Dearborn, 1997. 242 pp. (pbk. 0-57410-072-6) $22.95.
Gr. 9–12. This straightforward guide clearly explains how to market your invention and avoid common pitfalls in the process. The last chapter describes the ten ways to avoid invention suicide. There is a list by state of inventor assistance organizations and the author recommends that inventors contact one in their state. A bibliography and an index are included.

Pressman, David. *Patent It Yourself,* 9th ed. Berkeley, Calif.: Nolo, 2002. 512 pp. (pbk. 0-87337-801-6) $49.99.
Gr. 9–12. From invention to patent to marketing, this do-it-yourself guide contains a wealth of information and written by a patent attorney. Helpful websites and forms are also included. The book concludes with appendixes and an index.

Reynolds, Francis D. *Crackpot or Genius?: A Complete Guide to the Uncommon Art of Inventing.* Chicago: Chicago Review Press, 1993. 197 pp. (pbk. 1-5562-193-6) $14.95.
Gr. 9–12. Reynolds warns readers from the outset that few inventors ever make enough money from their inventions to support themselves. However, some people are natural inventors and continue to invent things even if they are not profitable. He based this book on a course he teaches on how to invent. It contains useful ideas, practical suggestions, and vignettes about inventors and their inventions. A bibliography and an index are included.

Stim, Richard. *License Your Invention: Sell Your Idea and Protect Your Rights with a Solid Contract*, 3rd. ed. Berkeley, Calif.: Nolo, 2002. 352 pp. (pbk. 0-87337-857-1) $39.99.
Gr. 9–12. This book is for inventors who want to license their inventions to others in order to have them market and sell the inventions. Charts, diagrams, and sidebars contain additional information. A CD-ROM of forms, an appendix of forms, and an index are included.

Stim, Richard, and David Pressman. *Patent Pending in Twenty-Four Hours*. Berkeley, Calif.: Nolo, 2002. 250 pp. (pbk. 0-87337-793-1) $29.99.
Gr. 9–12. A provisional patent application (PPA) establishes a claim to the invention before the patent application is filed, thus allowing the inventor to use the term "patent pending." This book explains how to complete a PPA, contains the essential forms, and provides instruction on completing the forms. Six appendixes conclude the book.

Vaidhyanathan, Siva. *Copyrights and Copywrongs: The Rise of Intellectual Property and How It Threatens Creativity*. New York: New York University Press, 2001. 243 pp. (hc. 0-8147-8806-8) $40.00.
Gr. 9–12. This book examines copyright law from the perspective that it is insensitive to the cultural values of some populations and that is stifles creativity. Students will be particularly interested in the portions of the book that deal with music copyright infringements. Notes and an index conclude the book.

Warda, Mark. *How to Register Your Own Copyright*, 4th ed. Naperville, Ill.: Sphinx, 2002. 213 pp. (pbk. 1-57248-200-1) $24.95.
Gr. 9–12. The book contains a great deal of information on copyrights, including among other things what a copyright is, how to obtain one, and the benefits of obtaining one. The book is well organized, which makes it easy for readers to find the information they need. A glossary, appendixes, and an index are included.

——. *How to Register Your Own Trademark*, 3rd. ed. Naperville, Ill.: Sphinx, 2000. 225 pp. (pbk. 1-57248-104-8) $21.95.
Gr. 9–12. An introduction to trademarks provides readers the background information they need to comprehend the material in the text. From designing to registering a trademark, this book provides a comprehensive plan. Appendixes, a bibliography, and an index conclude the book.

Battle, Carl. *The Patent Guide: A Friendly Guide to Protecting and Profiting from Patents*. New York: Allworth Press, 1997. 209 pp. (pbk. 1-880559-72-2) $18.95.
Gr. 10–12. Once you have an idea for an invention, this book has the answers to questions about obtaining a patent and copies of forms to complete. It is filled with useful information including how to deal with invention brokers, promotion firms, patent attorneys, and patent agents. Appendixes and an index are included.

Crews, Kenneth D. *Copyright Essentials for Librarians and Educators*. Chicago: American Library Association, 2000. 143 pp. (pbk. 0-8389-0797-0) $45.00.
Gr. 10–12. Sections in the book include formalities and duration; owners and rights; fair use; special exceptions; and looking ahead. The book addresses the fundamentals of copyright law in an easy-to-understand manner. This slim volume has a well-designed layout that makes it easy to find the information. Seven appendixes conclude the book.

Fishman, Stephen. *The Copyright Handbook: How to Protect and Use Written Works*. Berkeley, Calif.: Nolo, 2002. 432 pp. (pbk. 0-87337-855-5) $39.99.
Gr. 10–12. While the intent of the book is to help people file for copyright protection and protect their copyright, the book also contains information on fair use that is of benefit to educators. Appendixes of forms, which are also available on a CD-ROM in the back of the book, and an index are provided.

Grissom, Fred, and David Pressman. *The Inventor's Notebook*. Berkeley, Calif.: Nolo, 2000. 224 pp. (pbk. 1-87337-599-8) $24.99.
Gr. 10–12. The purpose of this notebook is to help inventors document their work in order to maintain a legal record and ultimately obtain a patent for their invention. Worksheets, forms, and a glossary are included.

Lo, Jack, and David Pressman. *The Patent Drawing Book*. Berkeley, Calif.: Nolo, 1997. 240 pp. (pbk. 1-87337-378-2) $29.95.
Gr. 10–12. The specifics of the formal drawings required for filing a patent are contained in this book. It includes information on drawing by hand and computer-aided drawing. Full-page line drawings, an appendix of forms, and an index are included.

Strong, William S. *The Copyright Book: A Practical Guide*, 5th ed. Cambridge, Mass.: The MIT Press, 1999. 376 pp. (hc. 0-262-19419-8) $34.95.

Gr. 10–12. The book explains legal issues regarding copyright law in an understandable language for those who are not lawyers. One chapter clearly explains the changes in the copyright law that became effective on January 1, 1978. Appendixes, notes, and an index are provided.

The Value of a Good Idea: Protecting Intellectual Property in an Information Economy. Los Angeles: Silver Lake, 2002. 436 pp. (pbk. 1-56343-745-7) $24.95.

Gr. 10–12. From the Taking Control series. Separate sections of the book contain information on copyrights, trademarks, patents, and other intellectual property. The purpose of the book is to provide readers with a working knowledge of what intellectual property is and how to protect their own. Appendixes and an index conclude the book.

Wherry, Timothy Lee. *Patent Searching for Librarians and Inventors*. Chicago: American Library Association, 1995. 89 pp. (pbk. 0-8389-0641-9) $25.00.

Gr. 10–12. This book begins with an introduction to patents and then describes how to conduct a patent search, which is an essential component of the patent application process. It is written in a clear, straightforward manner. The book includes a bibliography, appendixes, and an index.

Wilson, Linda Lee. *The Trademark Guide: A Friendly Guide to Protecting and Profiting from Trademarks*. New York: Allworth, 1998. 193 pp. (hc. 1-880559-81-1) $18.95.

Gr. 10–12. Wilson begins by making clear the distinctions between copyrights, patents, and trademarks. The book contains a through discussion of trademarks in concise, understandable text. Readers learn how not to infringe on others' trademarks and how to protect their own trademark. A concluding chapter discusses trademarks in cyberspace. Appendixes, a glossary, and an index conclude the book.

Appendix A

Professional Resources for Educators

This appendix contains book annotations, a listing of periodicals, and journal article annotations. The books below are resources for educators and some of them provide historical information about discoveries and inventions. In the periodicals section, there is subscription information on a collection of magazines for educators and students. These periodicals frequently carry articles on discoveries and inventions. In the last section of the appendix is a compilation of articles from journals and magazines that contain the latest information on a number of discoveries and inventions. Some of the items in this appendix will also be of interest and use to students.

Books

The book annotations in this section are arranged in alphabetical order by the author's last name. These books contain useful background information on discoveries and inventions. Some of the books look back at the history of technology and some offer cautions about the unexpected consequences of technology. Two of the books discuss the importance of creativity and offer ideas for fostering creativity.

Cardwell, Donald. *The Norton History of Technology*. New York: W.W. Norton, 1995. 563 pp. (hc. 0-393-03652-9) $35.00.
The book is organized chronologically into three sections: clockwork and Christianity; the Industrial Revolution; and power without wheels. This history of technology notes the impact of technology on society,

politics, and international cooperation. Notes, a short general bibliography, and an extensive index are included.

Chiles, James R. *Inviting Disaster: Lessons from the Edge of Technology*. New York: HarperCollins, 2001. 338 pp. (hc. 0-06-662081-3) $15.95.
When man and machines collide, the consequences can be disastrous. Chiles asserts that with today's "smart systems" it takes only a minor mishap to cause a tragedy and the most vulnerable area seems to be aviation. A list of the disasters, calamities, and near misses cited in the book, a list of key sources, and an index are provided.

Crouch, Tom D. *Wings: A History of Aviation from Kites to the Wright Brothers to the Space Age*. New York: W.W. Norton, 2003. 512 pp. (hc. 0-393-05767-4) $29.95; (pbk. 0-393-32227-0) $14.95.
The successes and failures of aviation pioneers are chronicled in this fascinating narrative of exhilarating flights and moments of sheer terror. Crouch captures the excitement and the courage of these pioneers and one hundred and twenty-five illustrations depict the pioneers and their flying machines.

Csikszentmihalyi, Mihaly. *Creativity: Flow and the Psychology of Discovery and Invention*. New York: HarperCollins, 1996. 456 pp. (hc. 0-06-017133-2) $27.50; (pbk. 0-06-092820-4) $15.00.
Based on the author's research that involved interviewing one hundred exceptional individuals, he explores the creative process and explains why it should be cultivated. Appendixes, notes, references, and an index are included.

Gelb, Michael J. *How to Think like Leonardo da Vinci: Seven Steps to Genius Every Day*. New York: Delacorte Press, 1998. 322 pp. (hc. 0-385-32381-6) $24.95.
Following an introduction to da Vinci, Gelb describes how to develop seven principles in order to foster creative thinking. Activities and exercises are included, many of which can be adapted to the classroom. A chronology, a list of books for recommended reading, resources, and a list of illustrations are included.

Hamza, Khidhir, with Jeff Stein. *Saddam's Bombmaker: The Terrifying Inside Story of the Iraqi Nuclear and Biological Weapons Agenda*. New York: Scribner, 2000. 352 pp. (hc. 0-684-87386-9) $26.00.

Hamza was the former head of Saddam Hussein's nuclear program. Hamza describes his work for Hussein and his harrowing escape from Iraq. This riveting book leaves readers wondering about the bomb-making programs of other countries. An epilogue and an index conclude the book.

Lienhard, John H. *Inventing Modern: Growing Up with X-Rays, Sky-scrapers, and Tailfins*. New York: Oxford University Press, 2003. 292 pp. (hc. 0-19-516032-0) $28.00.
This book is a nostalgic look at things once considered modern. Lienhard examines the technological advances of the nineteenth century as they changed his grandfather's life from the years 1822 to 1903. He discusses the modern advances during his own life and proposes the word "expanded" to describe the technology his great-grandchildren will encounter. Black and white photographs, reproductions, notes, and an index are provided.

Loewry, Raymond. *Never Leave Well Enough Alone*. Baltimore, Md.: Johns Hopkins University Press, 2002, 1951, 1950. 377 pp. (hc. 0-8018-7211-1) $29.95.
Born in France, industrial designer Raymond Loewry moved to the United States when he was twenty-five years old. His amusing, instructive autobiography describes how he redesigned American consumer goods to make them more appealing. His successful redesigns include the Exxon symbol, the Lucky Strike package, and Sears refrigerators. Black and white photographs of some of his designs are included.

Muir, John. *Nature Writings*. New York: The Library of America, 1997. 888 pp. (hc. 1-883011-24-8) $35.00.
This single volume includes the most important and significant works of John Muir, a crucial figure in the creation of our national parks system and a prophet of environmental awareness. From his early years in Scotland to his beloved California, John loved the wilderness and wanted to preserve it for future generations.

Pacey, Arnold. *Meaning in Technology*. Cambridge, Mass.: The MIT Press, 1999. 264 pp. (hc. 0-262-16182-6) $35.00.
At times it seems as though technology is being forced upon us with little regard to our feelings or needs. Pacey shows how the personal experience of technology shapes and changes technology. He also dis-

cusses the importance of the context in which the technology is found and the last chapter discusses people-centered technology. Notes and an index conclude the book.

——. *Technology in World Civilization: A Thousand-Year History.* Cambridge, Mass.: The MIT Press, 1990. 238 pp. (hc. 0-262-16117-6) $30.00.

Pacey discusses how similar inventions appear in the world in different places and at different times with seemingly no connections between the developments. At other times connections can be made between how inventions were developed. He also discusses the transfer of technology when an invention moves from one country to another. One example of this transfer is the construction of nuclear power plants in different parts of the world. Notes, a bibliography, and an extensive index complete the book.

Rhodes, Richard. *Dark Sun: The Making of the Hydrogen Bomb.* New York: Simon & Schuster, 1995. 731 pp. (hc. 0-684-80400-X) $32.50; (pbk. 0-684-82414-0) $18.00.

The meticulous research by this Pulitzer Prize–winning author resulted in a thorough examination of the United States' and Soviet Union's race to create the hydrogen bomb. Rhodes documents the politics and espionage of the Cold War that resulted in a stalemate. Black and white photographs, notes, an extensive bibliography, and an index are included.

——. *Making of the Atomic Bomb.* New York: Simon & Schuster, 1995. 747 pp. (pbk. 0-684-81378-5) $20.00.

Rhodes won the Pulitzer Prize in 1988 for this engrossing book. The book contains biographies of the key individuals involved in the Manhattan Project and detailed explanations of the technologies used in the construction of the bomb. Black and white photographs, notes, an extensive bibliography, and an index are included.

Root-Bernstein, Robert, and Michele Root-Bernstein. *Sparks of Genius: The Thirteen Thinking Tools of the World's Most Creative People.* Boston: Houghton Mifflin, 1999. 401 pp. (hc. 0-395-90771-3) $26.00.

The book begins with a chapter titled "rethinking thinking" and ends with a chapter titled "synthesizing education." The chapters in between describe ways to enhance thinking such as abstracting, body thinking, dimensional thinking, and transforming. Notes, a bibliography, a list of minds-on resources, and an index are included.

Shapin, Steven. *The Scientific Revolution*. Chicago: University of Chicago Press, 1996. 218 pp. (hc.0-266-75020-5) $19.95; (pbk. 0-266-75021-3) $12.00.
Shapin examines the what, how, and why of the Scientific Revolution in three chapters. By examining all three of these questions, he provides a well-rounded examination of science in the seventeenth century. Black and white illustrations, a bibliographic essay, and an index are included.

Tenner, Edward. *Our Own Devices: The Past and Future of Body Technology*. New York: Alfred A. Knopf, 2003. 314 pp. (hc. 0-375-40722-7) $26.00.
This is an enlightening examination of body technology inventions that are used in ways never intended by their inventors and those inventions that have produced unintended consequences. For example, helmets first introduced in World War I are now found in on coalminers, on construction workers, and on cyclists. An epilogue, notes, suggestions for further reading, and index are provided.

———. *Why Things Bite Back: Technology and the Revenge of Unintended Consequences*. New York: Alfred A. Knopf, 1996. 346 pp. (hc. 0-679-42563-2) $26.00.
This book is a compelling look at the "revenge effects" of the technological advances that have made lives easier. Tenner cautions readers to think about and anticipate these effects before adopting new technologies. Books for further reading, notes, and an index are included.

Teresi, Dick. *Lost Discoveries: The Ancient Roots of Modern Science–From the Babylonians to the Maya*. New York: Simon & Schuster, 2002. 453 pp. (hc. 0-684-83718-8) $27.00.
Teresi takes readers on a multicultural trip back in time to show how the discoveries of ancient non-Western scientists and scholars have been overlooked. Discoveries are organized by chapters including mathematics, astronomy, cosmology, physics, geology, chemistry, and technology. The book contains extensive notes and an extensive bibliography, both organized by chapters. An index is provided.

Uglow, Jenny. *The Lunar Men: Five Friends Whose Curiosity Changed the World*. New York: Farrar, Straus and Giroux, 2002. 588 pp. (hc. 0-374-19440-8) $30.00.
Matthew Boulton, James Watt, Josiah Wedgewood, Erasmus Darwin, and Joseph Priestley were the core members of a distinguished group that met each full moon and called themselves the Lunar Society of Birmingham. Other members of the group included: James Keir, William Small, William Withering, Richard Lovel Edgeworth, and Thomas Day. The ideas of this group of men led the way to the Industrial Revolution. An epilogue, a chronology, sources, and an index conclude the book.

Periodicals

Listed below are periodicals that frequently contain articles on discoveries and inventions. The periodicals are presented in alphabetical order. Subscription information is provided, as are websites for the periodicals. Some of the organizations listed in Appendix C offer periodicals to their members.

American Heritage
Syracuse, N.Y.: Forbes
http://www.americanheritage.com/index.shtml
Bimonthly ($12)

American Scientist
Research Triangle Park, N.C.: Sigma Xi, The Scientific Research Society
http://www.americanscientist.org/
Bimonthly ($85 Inst./$28 Ind.)

Ask: Science and Discovery
LaSalle, Ill.: Cobblestone Publishing
http://www.cobblestonepub.com/
Nine issues per year ($32.97)

Astronomy
Wasukesha, Wis: Kalmbach Publishing
http://www.astronomy.com/homc.asp
Monthly ($39.95)

Aviation Week and Space Technology
New York: The McGraw-Hill Companies
http://www.aviationnow.com
Fifty-one issues per year ($98)

Current Science
Stamford, Conn.: Weekly Reader Corporation
http://www.weeklyreader.com/teens/current_science/
Sixteen issues per year ($9.75)

Discover: Science, Technology, and Medicine
Buena Vista, Fla.: Buena Vista Magazines
http://www.discover.com
Monthly ($24.95)

The Futurist
Bethesda, Md.: World Future Society
http://www.wfs.org/
Bimonthly ($55 Inst./$45 Ind.)

The International Journal of Robotics Research
Cambridge, Mass.: Sage Publications
http://www.ijrr.org/about.html
Monthly ($995 Inst./$150 Ind.)

Inventor's Digest
Boston, Mass.: Affiliated Inventors Foundation
http://www.inventorsdigest.com/
Monthly ($27)

Muse: Science and Discovery
LaSalle, Ill.: Cobblestone Publishing
http://www.cobblestonepub.com/
Nine issues per year ($32.97)

National Geographic
Washington, D.C.: National Geographic Society
http://www.nationalgeographic.com/
Monthly ($34)

National Geographic Explorer
Washington, D.C.: National Geographic Society
http://www.nationalgeographic.com/
Six issues per year ($4.95 per student)

National Geographic for Kids
Washington, D.C.: National Geographic Society
http://www.nationalgeographic.com/
Ten issues per year ($17.95)

New Scientist
London: Reed Business Information
http://www.newscientist.com/
Fifty-one issues per year ($89)

Odyssey: Adventures in Science
LaSalle, Ill.: Cobblestone Publishing
http://www.cobblestonepub.com/
Nine issues per year ($29.95)

Popular Mechanics
New York: Hearst Communications
http://popularmechanics.com/
Monthly ($24)

Popular Science
Boulder, Colo.: Time4 Media
http://www.popsci.com/popsci/
Monthly ($12.95)

Science News: The Weekly Newsmagazine of Science
Marion, Ohio: Science Service
http://www.sciencenews.org/
Fifty-one issues per year ($54.50)

Science Weekly
Silver Spring, Md.: 20904
http://www.scienceweekly.com/
Sixteen times per year from September to May ($4.95 per student/$14.95 Ind.)

Scientific American
New York: Scientific American
http://www.sciam.com/
Monthly ($34.97)

Sky and Telescope
Belmont, Mass.: Sky Publishing Corp
http://skyandtelescope.com/
Monthly ($42.95)

Smithsonian
Washington, D.C.: Smithsonian Institution, 1969–
http://www.smithsonianmag.si.edu/
Monthly ($29)

Technology Review
Cambridge, Mass.: Massachusetts Institute of Technology
http://www.technologyreview.com/
Ten issues per year ($28)

Articles

The articles annotated below include a broad survey of discoveries, inventions, and current research in a variety of areas. These articles show that a wide range of magazines, journals, and newspapers contain stories about discoveries and inventions. The articles annotated below are arranged in alphabetical order by the author's last name.

Alexander, Max. "Wow, Isn't that Cool!" *Smithsonian* 34, no. 6 (Sept. 2003): 95–96.
In this interview, inventor Dean Kamen talks about the often messy, failure-prone invention process. Kamen invented the Segway Human Transporter and the iBot wheelchair.

Ashley, Steven. "Alloy by Design." *Scientific American* 289, no. 1 (July 2003): p. 24.
Researchers designed Toyota's titanium-based alloys while seated in front of a computer screen using computational models. Then, they went into the lab to create the alloys. Using the computer to create al-

loys enabled the researchers to bypass traditional trial-and-error methods.

Banchereau, Jacques. "The Long Arm of the Immune System." *Scientific American* 287, no. 5 (Nov. 2002): 52–59.
Ongoing research is discovering ways to harness the therapeutic potential of dendritic cells, which are found in body tissues. A two-page color diagram explains how dendritic cells fight infection and a chart shows dendritic cell cancer vaccines that are under development.

"A Bugs Life for Robots." *Economist* 366, no. 8315 (Mar. 15, 2003): 10–11.
In the world of robots, bugs are crawling all over. Observing the movements of cockroaches, geckos, snakes, spiders, and crabs has enabled scientists to construct robots with legs that enable them to travel in places robots with wheels cannot.

Baxter, Roberta. "The Golden Age of Invention and Innovation." *Cobblestone* 21, no. 4 (Apr. 2000): 19–24.
A brief overview of inventions from the nineteenth-century is the focus of this article. A sidebar describes the World's Colombian Exposition of 1893 (World's Fair). A short glossary is included.

Chowder, Ken. "Eureka!" *Smithsonian* 34, no. 6 (Sept. 2003): 92–94, 97.
Accident and serendipity are at times a part of the invention process. Chowder describes several examples in this article including Fleming's accidental discovery of the role of penicillin and the invention of Teflon. The article makes the point that it takes an astute observer and a creative thinker to recognize and develop accidental or serendipitous occurrences.

Churchman, Chris. "Imagine the Possibilities!" *Science and Children* 39, no. 4 (Jan. 2002): 22–25.
Using potatoes, students proceed through the invention process. First they determine a need and then they design a way to meet that need using a potato. Students design their invention, diagram it, build it, and name it.

Curry, Andrew. "Taking Flight." *U.S. News and World Report* 135, no.
    2 (July 21, 2003): 38–46.
In commemoration of the one hundredth anniversary of the Wright
brothers' flight on December 17, 1903, this article includes a brief history of their struggles, information on their personal life, a visual timeline of milestones in flight, and an article on the next flight frontier.

D'Alto, Nick. "From the Notebooks of Leonardo." *Odyssey* 10, no. 8
    (Nov. 2001): 35–36.
Drawings from Leonardo's sketchbooks show a variety of inventions
including the contact lens and the rotisserie oven. Readers are encourage to try building one of Leonardo's inventions.

Daniels, James. "Helping Hand." *Boys' Life* 92, no. 10 (Oct. 2002): 12.
This is the story of Michael Schumann's invention, the Parkinson's
Glove. The glove has a top support bar and a wrist bar, which enables
people with Parkinson's disease to reduce tremors in their arms and
regain the use of their hands.

"Fishy Idea." *Current Science* 87, no. 8 (Dec. 7, 2001): 13.
The winners of the tenth to twelfth grade division of the Toshiba/NSTA
ExploraVision Science and Technology Competition invented the
aquagill. The aquagill is a suit lined with tubes filled with artificial he-
moglobin capable of absorbing oxygen from water and providing divers
with an unlimited air supply. This competition involves students in
designing inventions that will be useful twenty years into the future.

"Gecko Glue." *SuperScience* 14, no. 4 (Jan. 2003): 3.
The microscopic hairs on the bottom of geckos' feet enable them to
climb walls and move across glass. Using this inspiration from nature a
team of scientists created gecko glue, which is not a liquid like other
glues. It works more like a magnet rather than glue.

Gow, George. "Understanding and Teaching Creativity." *Tech Direc-
    tions* 59, no. 6 (Jan. 2000): 32–36.
Gow describes different types of creativity and describes mental exercises to improve creativity. He also discusses the importance of observation in fostering creativity.

Grim, Pamela. "Too Close to Ebola." *Discover* 24, no. 6 (June 2003):
    42–47.

Working for the summer in Gulu, Uganda an American doctor encountered the Ebola virus. She describes the horrors of the disease and the fact that nurses and doctors treating the disease are likely to contract it and die. Sidebars explain how Ebola works and present a brief history of the disease.

Hall, Stephen S. "On the Trail of the West Nile Virus." *Smithsonian* 34, no. 4 (July 2003): 88–102.
As the West Nile Virus spreads across the United States and around the globe, some researchers are working on a vaccine while other researchers are studying the long-term effects of this infection. The author compares the outbreak of West Nile Virus to other mosquito-borne infections such as yellow fever and St. Louis encephalitis.

Haralson, Darryl. "Basketball Buddies Build Robot Shot Doctor." *USA Today* (Feb. 7, 2003): 01b.
Three Silicon Valley basketball buddies invented a "seeing" computer to help basketball players improve their shots. The article describes how the robot works and that it is being tested by basketball players at the national, university, and high school levels.

Harbison, Martha. "A Brief, Recent Flash in the History of Photography" *Popular Science* 263, no. 2 (Aug. 2003): 54–56.
This timeline uses text and photographs to show the evolution of the camera and includes predictions on what the future of photography holds.

Hendricks, Mark. "Spin Control." *National Geographic World* 307 (Mar. 2001): 14–18.
Innovations have given the world's oldest toy, the yo-yo, a new spin that enables it to do a variety of surprising tricks. This article describes the innovations, introduces young yo-yo enthusiasts, and describes their tricks. Step-by-step instructions for "The Sleeper" trick are included.

"History's Top Invention." *Science World* 59, no. 12 (Mar. 28, 2003): 7.
A survey of one thousand four hundred Americans determined that the most indispensable invention is the toothbrush. Teenagers' top five most indispensable inventions are the toothbrush, the automobile, the computer, the cell phone, and the microwave.

Krebs, Danute V., and Barbara D. Clark. "Camp Invention Connects to Classrooms." *Gifted Child Today Magazine* 23, no. 3 (May/June 2000): 28–34.
An overview of Camp Invention, how it impacts gifted children, and details of the curriculum modules are included in this article.

Langreth, Robert. "The Doctor Is In." *Forbes* 172, no. 4 (Sept. 1, 2003): 88–89.
This brief article reports on monitors implanted in patients. They have the potential to help victims of chronic diseases such as diabetes and heart failure by monitoring the patients' vital statistics and transmitting the information directly to doctors in their offices.

Lutton, Louise Pietsch. "Mechanics and Inventors." *Arts and Activities* 124, no. 3 (Nov. 1998): 36.
Beginning with a collection of broken small appliances a first grade teacher integrated science and art into his classroom curriculum. After reviewing simple machines with the students, they took apart the appliances searching for the simple machines within them. The next week the students drew an invention that would make their lives easier.

Mauch, Elizabeth. "Using Technological Innovation to Improve the Problem-Solving Skills of Middle School Students." *Clearing House* 74, no. 4 (Mar./Apr. 2001): 211–214.
This article reports on a grant that provided funding for a hands-on workshop to show middle school teachers how to use Lego Mindstorms Robotic Invention Systems to improve the problem-solving skills of their students. Descriptions of the benefits to the students and the challenges faced by the teachers help educators decide if this program would be useful for their students.

McCollum, Sean. "Ten Cool Inventions." *National Geographic for Kids* 330 (May 2003): 33–36.
McCollum describes ten cool inventions and tells how they work. These are inventions that appeal to children. Included are eyeglasses that project a computer screen in front of the wearer's eyes; a microchip that is implanted in a molar by which wearers receive phone calls directly in their ear; and a jacket with a built-in sound system.

"Not So Green?" *Science News* 163, no. 24 (June 14, 2003): 373.
Hydrogen fuel may not be as environmentally friendly as its proponents argue. Leaks of hydrogen fuel from production facilities, storage facilities, and fuel cells that use it could potentially harm the atmosphere.

Plucker, Jonathan A. "What's in a Name? Young Adolescents' Implicit Conceptions of Invention." *Science Education* 88, no. 2 (Mar. 2002): 149–160.
Fifty-five sixth grade students responded to open-ended survey questions to determine their conceptions of "invention." The results of the survey indicated that students do not fully appreciate the value of reflection in the invention process. Further, the author contends that rather than short-term cookbook approaches to the study of inventions, students need in-depth project-based or problem-based learning to fully understand and appreciate the invention process.

"Prototype." *Technology Review* 106, no. 4 (May 2003): 14–16.
This column briefly describes eight new inventions including a new method for creating x-rays, a smart wet suit, and a new procedure to make bone-marrow transplants more successful.

Reed, J.D. "Virus Vanquisher." *Smithsonian* 32, no. 11 (Feb. 2002): 28–29.
D. A. Henderson led the World Health Organization's (WHO) campaign to eradicate smallpox. With threats of bioterrorism he is once again ready to lead the fight against smallpox.

Rosenberg, Barry. "NASA Technology Sharpens Images with Jitters." *Aviation Week and Space Technology* 158, no. 20 (May 19, 2003): 61.
Video Image Stabilization and Registration (VISAR) was selected as NASA's Commercial Invention of the Year. VISAR, a computer algorithm, sharpens video images. It was used to sharpen images of the *Columbia* space shuttle disaster and is now being used by law enforcement and military agencies.

Scanlon, Lisa. "Behind Bars." *Technology Review* 106, no. 3 (Apr. 2003): 80.
In 1952 inventors, Norman Joseph Woodland and Bernard Silver received a patent for a linear bar-code system. More than twenty years later their invention was put to use by the grocery industry.

——. "Good Vibrations." *Technology Review* 106, no. 2 (Mar. 2003):
    80.
The aetherphone was invented in 1920 by a Soviet physicist and cello
player, Leon Theremin. This musical instrument produces tunes without being touched. The article briefly describes the history of the instrument and its recent revival.

Sheppard, Robert. "Medicine in 2020." *Maclean's* 113, no. 3 (Jan. 10,
    2000): 40–43.
Step into the future and discover what new medical treatments are on
the horizon. The scenario that unfolds is based on medical advancements currently available. A seventy-year-old woman and her son who
is hospitalized after an accident are the characters in this medical won-
derland.

Shulman, Seth. "The Vision Thing." *Technology Review* 108, no. 4
    (May 2003): 75.
This article discusses two inventions that will make it possible for people in the developing world to have affordable easy-to-obtain eyeglasses. Shulman makes the point that these inventions are being developed and marketed by individuals, rather than large corporations.

Smith, Elliot Blair. "USMC Ingenuity Marches On." *USA Today* (Mar.
    18, 2003): 07d.
Smith reports on a variety of inventions created by U.S. Marines sta-
tioned in Iraq. The inventions include among others a desert bunker, an
easy chair, a chin-up bar, and a hand-held washing machine. These
resourceful Marines scavenged the area for materials to create their
inventions.

"Speed Reading the Book of Life." *Economist* 367, no. 8329 (June 21,
    2003): 13.
The "DNA Prism" enables researchers to sort DNA segments in seconds rather than days. This short article describes how the prism works
and the benefits it offers to researchers working on gene sequencing.

Stone, Richard. "Championing a Seventeenth-century Underdog." *Sci-
    ence Now* (July 8, 2003): 1–2.
Robert Hooke is beginning to get the recognition he deserves after being hidden in Isaac Newton's shadow according to Stone. Hooke first

proposed a theory of evolution two centuries before Darwin. He for-
mulated Hooke's law that stress is directly related to strain.

"Students' Prizewinning Idea May Give Sight to Sore Eyes" *Current
     Science* 87, no. 9 (Dec. 21, 2001): 14.
A Photo-Electric Eye Prosthesis (P.E.E.P.) is an implant that would
enable visually impaired people to see. This was one of the winners in
the 2001 Toshiba/NSTA ExploraVision Science and Technology Competition.

"Sweet Freeze." *Current Science* 88, no. 12 (Feb.7, 2003): 8–9.
This article explains how ice cream is made and describes the research
of food scientist Douglas Goff, who has discovered a protein in winter
wheat that when added to ice cream keeps the ice crystals small resulting in smoother, tastier ice cream.

Tenner, Edward. "YOU Bought It. WHO Controls It." *Technology Re-
     view* 106, no. 5 (June 2003): 61–65.
Students familiar with the lawsuit against Napster or unable to make
copies of a favorite CD eagerly join in the debate about copyright protection presented in this article.

Thieme, Trevor. "The DOE's Gadget Guru." *Popular Science* 262, no.
     4 (Apr. 2003): 92.
Fariborz Bzorgi works at the Y–12 National Security Complex where
along with two thousand five hundred other scientists and craftsmen,
they invent items for the government and industry. The article briefly
describes some of Bzorgi's inventions such as the Hospital in a Box.
Illustrations accompany the article.

Trachtman, Paul. "Hero for Our Time: Louis Pasteur and Anthrax."
     Smithsonian 32, no. 10 (Jan. 2002): 34–41.
French veterinarian Monsieur H. Rossignol challenged Pasteur to prove
that his vaccine would prevent anthrax in sheep, which it did. From
Pasteur's extensive research came the disciplines of immunology and
bacteriology along with vaccines to protect the world. The importance
of Pasteur's work and his legacy are being remembered since anthrax
was spread through the U.S. mail.

"Watch this Space." *New Scientist* 177, no. 2376 (Jan. 4, 2003): 24–30.
Eleven technology milestones to watch for are profiled in this article.
From iron rice to stem cell banking to the first human clone, this is an
impressive list of possible milestones.

"Wearable Computer Wins Student $20,000." *Current Science* 86, no.
    4 (Oct. 20, 2000): 14.
Eighteen-year-old Steve Cosenza's wearable computer won him the
Duracell/National Science Teachers Association's Invention Challenge.
His invention includes a one-hand mobile keyboard composed of six
rings that when pressed together in different combinations operate the
computer.

Wilson, Jim. "No More Needles." *Popular Mechanics* 180, no. 6 (June
    2003): 42–44.
Researchers are developing ways to deliver drugs directly into the body
without needles. The transdermal drug delivery (TDD) system involves
painting the drug on a section of the skin and then bombarding the skin
with ultrasound, which causes the drug to penetrate the skin and seep
into the body. Another method under development uses effervescent
compounds that are taken orally.

Appendix B

Electronic Resources

The electronic resources listed in this appendix include software, videos, DVDs, and Internet websites. Approximate grade levels are given for the software and videos, which are arranged alphabetically. The websites listed in this appendix were active at the time of publication.

Software

The software described below actively involves students in learning science concepts, discovering how things work, and meeting famous scientists.

Discoveries: Explore Aviation and Flight from Sunburst Technology takes students on a journey of discovery as they learn the history of aviation and discover the mechanics of flight. As they travel they record their adventures in an on-screen journal. There is a game center and opportunities for independent explorations. This software is appropriate for grades three to six and is available on CD-ROM for both Windows and Macintosh computers.

Inventor Labs: Technology from Sunburst Technology takes students into the laboratories of Thomas Edison, Alexander Graham Bell, and James Watt. Students can explore the inventors' inventions as they learn about the history and science involved in the inventions. Films, photographs, documents, and audio files are included in the program.

This software is appropriate for grades five to twelve and is available on CD-ROM for both Windows and Macintosh computers.

Inventor Labs: Transportation from Sunburst Technology introduces students to the Wright brothers and their airplane; George Stephenson and his Rocket Train; and Gottieb Daimler and his 1901 Mercedes Simplex. Students can conduct experiments in a wind tunnel or design and experiment with racecars. The program contains photographs, copies of patents, and reproductions of historic documents. This software is appropriate for grades five to twelve and is available on CD-ROM for both Windows and Macintosh computers.

Learn about Physical Science: Simple Machines from Sunburst Technology enables students to discover how simple machines work. There is a matching game and a sorting game. A writing activity is included where students fix broken toys and then write about how they fixed the toys. A teacher's guide is available. This software is appropriate for kindergarten to second grade and is available on CD-ROM for both Windows and Macintosh computers.

The New Way Things Work 2.0 from DK Multimedia is based on David Macaulay's book by the same name. Animations, text, videos, and online resources are available to students as they learn the history of machines, meet the inventors, explore scientific principles, and discover how things work. This software is appropriate for grades three to twelve and is available on CD-ROM for both Windows and Macintosh computers.

Pinball Science from DK Multimedia has students examining an inventor's log to learn about science. They design and build pinball games, and as their expertise increases, more gadgets become available to use in their designs. This software is appropriate for grades three to eight and is available on CD-ROM for both Windows and Macintosh computers.

The Scientific Revolution by CLEARVUE explores scientific advances from the late Middle Ages to the Age of Reason. Students learn about the key ideas, discoveries, and inventions made during this period. This software is appropriate for grades seven to twelve and is available on CD-ROM for both Windows and Macintosh computers.

Thinkin' Science from Sunburst Technology involves students in using the scientific method as they conduct experiments in earth science, life science, and physical science. As students work on the software program, the experiments become more difficult. This software is appropriate for kindergarten to second grade and is available on CD-ROM for both Windows and Macintosh computers.

What's the Big Idea, Ben Franklin? by Scholastic is based Jean Fritz's book by the same name. The author narrates the story. The CD contains video clips, a glossary, and games. This software is appropriate for grades one to three and is available on CD-ROM for both Windows and Macintosh computers.

Videos

Below are brief descriptions of videos that can be used to help students learn about discoveries and inventions. The descriptions contain information on the length of the video and appropriate grade levels. Many of the videos come with a teacher's guide. These guides contain materials to help the teacher or librarian introduce the video to students and they contain ideas for extending the video.

Ancient Inventions from A and E Television Networks travels back in time to explore the construction of the pyramids and to examine Leonardo da Vinci's sketches for modern inventions. This fifty-minute video is appropriate for grades six to twelve.

Alexander Graham Bell: Voice of Invention from A and E Television Networks introduces Bell and some of his more famous inventions. Viewers learn that he was one of the original members of the National Geographic Society and was a lifelong friend of Helen Keller. This fifty-minute video is appropriate for grades six to twelve.

Building Big with David Macaulay: Educational Set from WGBH Boston examines what it takes to build large structures including bridges, dams, tunnels, skyscrapers, and domes. A teacher's guide is included. This set consists of six one-hour videos and is appropriate for grades four to eight.

Breaking the Code is an ExxonMobil Masterpiece Theatre and is available from WGBH Boston. It tells the story of Alan Turing's private life and the computer he designed to break the German Enigma code. This ninety-minute video is appropriate for grades nine to twelve.

Cannons from A and E Television Networks chronicles the evolution of cannons from the thirteenth century to the present. This fifty-minute video is appropriate for grades six to twelve.

Captured Light: The Invention of Still Photography from A and E Television Networks traces the evolution of the still camera and introduces the men who developed still photography. Photographers introduced in the video include Joseph-Nicéphore Niepce, Louis-Jacques-Mandé Daguerre, William Henry Fox Talbot, and George Eastman. This fifty-minute video is appropriate for grades six to twelve.

Clocks from A and E Television Networks explores the world of time and how it has been measured from water clocks to atomic clocks. This fifty-minute video is appropriate for grades six to twelve.

The Creation of the Computer from A and E Television Networks looks at the history of the computer from Charles Babbage's calculating machine to the IBM punch cards to the computers of today. This fifty-minute video is appropriate for grades six to twelve.

Marie Curie More Than Meets the Eye from Devine Entertainment is set in Paris during World War I. This film depicts Marie Curie and her daughter Irene as they transport x-ray equipment to hospitals on the front lines. Two young sisters who live nearby begin watching the quiet Madame Curie and decide she must be a spy. When they discover her glowing laboratory, they become convinced she is a spy. This fifty-six-minute video is appropriate for grades three to eight.

The Edison Effect from A and E Television Networks is a three-volume set. Photographs and film clips introduce Edison and his trusted assistants as well as his rivals. This two and one half-hour video is appropriate for grades nine to twelve.

Edison: The Wizard of Light from Devine Entertainment features Edison and his kinetoscope, which was used to create motion pictures. When twelve-year-old Jack crashes into the Menlo Park laboratory,

Edison befriends him. Jack is by Edison's side as he develops the kinetoscope. This Emmy award–winning video was nominated for five Emmy awards. The fifty-six-minute video is appropriate for grades three to eight.

Einstein: Light to the Power of Two from Devine Entertainment is available on both video and DVD. Set in 1954 against the backdrop of the McCarthy Committee on Un-American Activities and blatant racism, Einstein meets a young African American girl, Lannie Willis. They forge a friendship as Einstein helps Lannie with her homework and encourages her to discover things for herself. Included on the DVD are a special feature on the life and times of Albert Einstein and the option of listening to the movie in English, French, or Spanish. The fifty-six-minute film is appropriate for grades three to eight.

Einstein Revealed from WGBH Boston paints a portrait of Einstein as a passionate young man, using material from his private notebooks and letters. Computer animations help viewers visualize the inner workings of Einstein's mind. This two-hour video is appropriate for grades seven to twelve.

Endangered Planet from WGBH Boston explores how the technological revolution has had harmful consequences for the environment. It describes the contamination of the environment and efforts to prevent further contamination. This one-hour video is appropriate for grades seven to twelve.

Fast Forward from WGBH Boston examines how new technologies bring benefits for some and hardships for others. New technologies are shrinking the world with a corresponding impact on their politics and economies. This one-hour video is appropriate for grades seven to twelve.

Ben Franklin: Citizen of the World from A and E Television Networks examines Franklin's inventiveness, his industriousness, and his many accomplishments. This fifty-minute video is appropriate for grades nine to twelve.

Galileo's Battle for the Heavens from WGBH Boston is based on Dava Sobel's *Galileo's Daughter: A Historical Memoir of Science, Faith, and Love*. Viewers learn about Galileo's discoveries, his inquisition trial, and life in the seventeenth century. This two-hour video is appropriate for grades seven to twelve.

Galileo: On the Shoulders of Giants is from Devine Entertainment and available on both video and DVD. This Emmy award–winning video introduces young viewers to Galileo's struggles to get his discoveries accepted. Viewers meet Galileo as he tutors a young boy of the Medici family. It was filmed on location in Italy and it includes authentic costumes that were created in Italy. Included on the DVD are interviews with the actors and the option of listening to the movie in three different languages. The fifty-six-minute video is appropriate for grades three to eight.

Great Inventions from A and E Television Networks takes viewers back in time to meet famous inventors and their inventions. The video also focuses on the impact of the inventions. This fifty-minute video is appropriate for grades nine to twelve.

Great Projects: The Building of America: The Big Dig from WGBH Boston explores the construction of an underground roadway in Boston. This one-hour video is appropriate for grades seven to twelve.

Leonardo: A Dream of Flight was produced by Devine Entertainment. Leonardo befriends Roberto, the crippled child of a bird seller at the Milan market. Intrigued by the glider hanging in the artist's studio Roberto and Leonardo's assistants take the glider into the countryside. Pushed by Leonardo's assistants, Roberto is airborne momentarily. The video was filmed on location near the northern city of Padua. This fifty-six-minute video is appropriate for grades three to eight.

Lost at Sea: The Search for Longitude from WGBH Boston tells the story of John Harrison, carpenter and clock maker, who solved the mystery of longitude. Richard Dreyfus narrates this one-hour video that is appropriate for grades seven to twelve.

A Man, A Plan, A Canal, Panama from WGBH Boston uses historic film footage and archival photographs narrated by author David McCullough to tell the story of the building of the Panama Canal. This one-hour video is appropriate for grades seven to twelve.

Masters of Technology from WGBH Boston introduces five revolutionary thinkers: Vinton Cerf, Donna Shirley, Geoffrey Ballard, Burt Rutan, and Robert Langer. Interviews with these five visionaries highlight their achievements and their failures. This two and one half-hour video is appropriate for grades seven to twelve.

Medical Imaging from A and E Television Networks explores the discovery of x-rays and their impact on medicine. The video traces the development of medical imaging through the years. This fifty-minute video is appropriate for grades nine to twelve.

Modern Marvels, Set One from A and E Television Networks is a four-volume set that examines these engineering feats: the Statue of Liberty, the Eiffel Tower, the Golden Gate Bridge, and Las Vegas. Each fifty-minute video is appropriate for grades six to twelve.

Modern Marvels, Set Two from A and E Television Networks is a four-volume set that examines these engineering feats: the Empire State Building, the Grand Coulee Dam, Mount Rushmore, and the Panama Canal. Each fifty-minute video is appropriate for grades six to twelve.

Modern Marvels, Set Three from A and E Television Networks is a five-volume set that examines construction highlights of: the Transcontinental Railroad, tunnels, gothic cathedrals, ocean liners, and domed stadiums. Each fifty-minute video is appropriate for grades six to twelve.

Sir Isaac Newton: The Gravity of Genius is from A and E Television Networks and highlights Newton's many discoveries. It also examines the far-reaching impact of his discoveries. This fifty-minute video is appropriate for grades nine to twelve.

Newton: A Tale of Two Isaacs from Devine Entertainment is narrated by a young boy named Isaac whose father was a protégé of the famous scientist. In this video Newton's formulation of his theories of celestial movement and gravity are explored, as is his personal life. This fifty-six-minute video is appropriate for grades three to eight.

J. Robert Oppenheimer: Father of the Atomic Bomb from A and E Television Networks includes footage of the Manhattan Project. Interviews with Oppenheimer's friends show his personal side, and his lawyer talks about Oppenheimer's ordeal during the McCarthy hearings. This fifty-minute video is appropriate for grades nine to twelve.

Rocket! from A and E Television Networks traces the evolution of rockets from ancient China to the Apollo program. The video introduces Robert Goddard and Werner Von Braun. This two-volume set is two hundred minutes and is appropriate for grades nine to twelve.

Steve Wozniak: The Wizard of the Apple from A and E Television Networks looks at the life of the man who invented the personal computer. Photos, film clips, and interviews help to tell his story. This fifty-minute video is appropriate for grades six to twelve.

Wilbur and Orville Wright: Dreams of Flying from A and E Television Networks explores the personal lives of these remarkable brothers and their struggles to prove that flight was possible. This fifty-minute video is appropriate for grades six to twelve.

What's the Big Idea, Ben Franklin? is a video based on the book by the same name written by Jean Fritz. Viewers observe how Franklin questioned his world and experimented to find answers. This thirty-minute video is appropriate for grades one to six.

Winding Your Way through DNA from the University of California is a series of three award-winning videos: *Stories from the Scientists, On Becoming a Scientist,* and *Promise and Perils of Biotechnology: Genetic Testing.* A thirty-two-page teacher resource book comes with the videos. This seventy-four-minute video is appropriate for grades nine to twelve.

Internet Sites

The Internet sites listed below contain a variety of resources for learning about discoveries and inventions. Additional websites can be found in Appendix D Directory of Museums and Organizations. Many of the museums and organizations have websites that include virtual tours, information on exhibits, and online resources for students and educators. The sites below were active at the time of publication.

About Goddard
http://www.gsfc.nasa.gov/indepth/about_drgoddard.html
The Goddard Space Flight Center hosts these pages about Robert Goddard's life and accomplishments.

Academy of Achievement
http://www.Achievement.org
Created in conjunction with Achievement Television, this website includes profiles, interviews, video clips, and audio clips featuring famous achievers. Curriculum materials are also available on the website.

The Astronomy Cafe
http://www.astronomycafe.net/
This award-winning site contains questions and answers, explanations of weird things, astronomy articles, and more.

Ask Dr. Teller
http://www.llnl.gov/ask-teller
Teller was a colleague of Albert Einstein, Niels Bohr, Werner Heisenberg, Enrico Fermi, and Robert Oppenheimer. This website includes responses to questions people have asked about his life.

Biography.com
http://www.biography.com
Type in the name of an inventor or discoverer on this website to view a brief biography of the individual. Sponsored by A and E Television Networks, this site has links to other resources.

Bioterror
http://www.pbs.org/wgbh/nova/bioterror/
This is a companion website to the film "Bioterror." Links on the site are for information about the history of biowarfare, future germ defenses, and making vaccines.

The Black Inventor Online Museum
http://www.blackinventor.com
This online museum contains a database of information about black inventors and their inventions. Some of the entries include illustrations and photographs.

Center for Bioethics
University of Pennsylvania
http://bioethics.net/hsbioethics/
The website includes an area with resources for high school students to help them with their bioethics homework and to learn more about this controversial topic.

Chemsoc
http://www.chemsoc.org/
The Royal Society of Chemistry maintains this site as a resource for chemists. There is a learning resources link for educators and students. The site also contains a visual periodic table.

The Discovery of Insulin
http://www.discoveryofinsulin.com
This comprehensive site includes biographies, a list of books, a list of movies, links to other resources, and Frederick Banting's Nobel lecture.

Doctor Over Time
http://www.pbs.org/wgbh/aso/tryit/doctor/
Try out this interactive website to discover how medical treatments have changed over the years. The site also contains links to information on the medical researchers noted for important medical breakthroughs, including Frederick Banting, Ernst Chan, and Dorothy Hodgkin.

Education World
http://www.education-world.com
This is a searchable database of lesson plans and Internet links on a variety of topics including discoveries and inventions.

EnchantedLearning.com: Zoom Inventors and Inventions
http://www.enchantedlearning.com/inventors/indexa.shtml
This database of inventors and inventions can be searched by name, by topic, or by country. Some of the brief entries contain hyperlinks to related information.

Energy Efficiency and Renewable Energy, United States Department of Energy
http://www.eere.energy.gov/
Resources for teachers and student include lesson plans, science projects, contest information, career information, teacher workshops, and

Dr. E's Energy Lab for elementary students. Some of the material in the Energy Lab is text based and suitable for older students.

EurekAlert
http://www.eurekalert.org/
Visit this website for the latest information on advances in scientific research. This website is maintained by the American Association for the Advancement of Science.

The Faces of Science: African Americans in Science
http://www.princeton.edu/~mcbrown/display/faces.html
The database on this website contains profiles of accomplished African American engineers and scientists.

Rube Goldberg: The Official Rube Goldberg Website
http://www.rube-goldberg.com/
Learn about the master of complicated, convoluted inventions that perform simple operations on this website. Rube Goldberg's name is synonymous with making simple things complicated. The site contains a brief biography of Rube Goldberg, information on the Rube Goldberg Machine Contest, and a gallery of his drawings.

The Great Buildings Collection
http://www.greatbuildings.com
Explore some of the greatest buildings in the world. Information on this website is arranged by buildings, by architects, and by locations. The site includes three-dimensional models.

The Great Idea Finder (TGIF)
http://www.ideafinder.com
This site celebrates innovation and innovators. Under the Idea History link students find Invention Facts and Myths, Inventor Profile, Innovation Timeline, and Invention Trivia Quiz. Idea Showcase contains information on outstanding improvements to products, processes, and services.

History of Kodak
http://www.kodak.com/US/en/corp/aboutKodak/kodakHistory/kodak History.shtml
This site includes a biography of George Eastman, information on his inventions, and a chronology of his company.

How Stuff Works

http://www.howstuffworks.com

Just as the title says, this site explains how things work. Links on the website are organized by these categories: computers, autos, electronics, science, home, money, travel, and people. Diagrams and photographs help explain how stuff works.

How Things Work

http://howthingswork.virginia.edu

Physics Professor Louis A. Bloomfield explains how things work on this website. The site contains a searchable database of information on physics and science in everyday life. There is a link to information about a home study course for high school physical science and physics teachers.

Hubble Site

http://hubblesite.org

The Space Telescope Science Institute maintains this website which features pictures from the Hubble Space Telescope. There are links to a news center, a gallery, discoveries, a guide to the telescope, and a reference desk.

International Space Station

http://spaceflight.nasa.gov/station/index.html

Find out what is happening aboard the International Space Station (ISS) and check to see when it will be visible from your backyard. Try out the ISS Interactives for a self-guided tour.

Invention and Design

http://jefferson.village.virginia.edu/~meg3c/id/id_home.html

Supported by several grants, this website has resources for teachers and students interested in invention and design including learning modules for hands-on explorations.

InventNet: Inventors Network

http://www.inventnet.com

From software to books to patent forms a variety of online resources for inventors are available on this website.

The Invention Connection in Cyberspace
http://www.inventionconnection.com
This online version of the Invention Convention is a networking service for inventors and investors. It is also a resource for learning about a wide variety of inventions.

Invention at Play
http://www.si.edu/lemelson/centerpieces/iap/
Sponsored by the Lemelson Center for the Study of Invention and Innovation at the Smithsonian National Museum of American History, this website explores the importance of play in the lives of inventors. There are links to games for students to play.

Inventive Kids
http://www.inventivekids.com
Students learn a great deal about inventions and inventors while playing the interactive games on this website.

InventorEd.org: Education and News for Inventors
http://www.inventored.org/
Inventor Ronald J. Riley maintains this page, which has an excellent collection of resources for inventive kids.

Inventors
http://inventors.about.com/
On this website there are links to information about famous inventions, famous inventors, minority inventors, conventions, magazines, organizations, patents, and lesson plans. A section of the website includes information for young inventors. The website has popup advertisements and flashing advertisements that are distracting.

Kids Cafe: Patent Café's Space for Young Inventors
http://kids.patentcafe.com/index.asp
Kids find information about inventors and resources to help them create their own inventions on this website. The site includes resources for educators and parents. It is sponsored by PatentCafe.com, which is a resource for inventors, patent attorneys, and intellectual property managers.

Kids Invent!
http://www.invention-engine.com
The Invention Engine for Kids walks them through the invention process and provides help along the way. The site includes resources for teachers and parents.

Kids Pages
http://www.uspto.gov/go/kids/
Sponsored by the United States Patent and Trademark Office this website includes information on contests, puzzles, games, and links to other resources for young inventors.

Lego
www.lego.com/eng/
The Lego website contains information on designing and building with Lego blocks. There are links to the Lego Club, a message board, and educator resources.

Lemelson–MIT Program
http://web.mit.edu/invent/index.html
This website sponsored by the Massachusetts Institute of Technology (MIT) celebrates inventors whose ideas have been turned into accomplishments. There are links to the inventor of the week, an inventor's handbook, games, and other resources.

MarcoPolo: Internet Content for the Classroom
http://www.marcopolo-education.org/
Using the search engine on the MarcoPolo website enables educators to simultaneously access educational resources on their partners' websites. The partners are *ARTSEDGE* sponsored by the John F. Kennedy Center for the Performing Arts; *EconEdLink* sponsored by the National Council on Economic Education; *EDSITEment* sponsored by the National Endowment for the Humanities; *Illuminations* sponsored by the National Council of Teachers of Mathematics; *ReadWriteThink* sponsored by the International Reading Association and the National Council of Teachers of English; *Science NetLinks* sponsored by the American Association for the Advancement of Science; and *Xpeditions* sponsored by the National Geographic Society.

MIT Artificial Intelligence Laboratory
http://www.ai.mit.edu
The goals of the lab are to develop an understanding of how the mind works, to develop an understanding of the nature of intelligence, and to create intelligent systems. Links provide information on current research and companies developed through spin-offs from research conducted in the laboratory.

Medicine through Time
http://www.bbc.co.uk/education/medicine/nonint/home.shtml
An interactive timeline, resources for teachers, and resources for pupils can be found on this website about medical breakthroughs and the researchers responsible for them.

MEDLINEplus Health Information
http://www.nlm.nih.gov/medlineplus
Search this collection for information on health, diseases, medicines, clinical trials, current health news, and other health-related issues. This site is also available in Spanish.

National Archives and Records Administration
http://www.archives.gov/welcome/index.html
A variety of online exhibits is available and the search engine links educators to resources for teaching about discoveries and inventions. The Digital Classroom link contains resources for educators.

National Center for Biotechnology Information
http://www.ncbi.nlm.nih.gov/
Sponsored by the National Library of Medicine and the National Institutes of Health, this website contains links to literature databases, molecular databases, an archive of life science journal articles, a guide to inherited diseases, and other resources.

National Human Research Institute
http://www.genome.gov/
Under educational resources on this website are genetic education modules and a talking glossary of genetic terms available in both English and Spanish.

NASA–National Aeronautics and Space Administration
http://www.nasa.gov
News and events, multimedia features, and missions can be found on
the site. There are links on the site for kids, students, and educators.

NASA Ask a Space Scientist
http://image.gsfc.nasa.gov/poetry/ask/askmag.html#list
Astronomer Sten Odenwald answers questions about the sun and the
earth on this website. It also contains an archive of previous questions
and answers.

NASA/Jet Propulsion Laboratory Robotics
http://robotics.jpl.nasa.gov
Meet NASA's robots and the people who developed them by clicking
on the links on these web pages. There are links to information on current research and in-depth descriptions of the work accomplished by
the robots.

Nobel e–Museum
http://www.nobel.se/
The website offers a searchable database of Nobel Laureates, games,
readings, recordings, and simulations.

Nuclear Fusion
http://www.pppl.gov/fusion_basics/pages/fusion_basics.html
The Princeton Plasma Physics lab hosts this site, which contains information on nuclear fusion, including the advantages of nuclear fusion
and how fusion power is produced.

Panama Canal History Museum
http://www.canalmuseum.com
Examine photographs and documents related to the building of the
Panama Canal in this virtual museum. There are stories to read and
links to related sites on this website.

Particle Accelerators
http://www2.slac.stanford.edu/vvc/accelerator.html
Learn about particle accelerators and how they work on this website.
Take a virtual tour of an atom smasher at this website hosted by the
Stanford Linear Accelerator Center.

People and Discoveries
http://www.pbs.org/wgbh/aso/databank/
This is a searchable database of people and discoveries from the twentieth century. The database is organized by people, discoveries, and topics. The five topics include medicine and health; physics and astronomy; human behavior; technology; and earth and life science.

Planetary Science Research Discoveries
http://www.psrd.hawaii.edu/index.html
NASA-sponsored scientists post their latest discoveries about meteorites, planets, and other solar system bodies on this website.

Stalking the Mysterious Microbe
http://www.microbe.org/
Be a sleuth and solve the microbe mysteries, read about microbes in the news, and try some experiments on this website.

Time 100: The Most Important People of the Twentieth-century
http://www.time.com/time/time100/
Click on one of the six categories: leaders and revolutionaries; artists and entertainers; builders and titans; scientists and thinkers; heroes and icons; and person of the century. Then, click on the name of an individual to access a brief biography and information on why the person was selected one of the most important people of the twentieth-century.

Time Travel
http://www.pbs.org/wgbh/nova/time/
This website is a companion to the NOVA program that examined whether or not time travel is possible. There are links on the site to Carl Sagan and Albert Einstein's theories on time travel as well as other links to information on time travel. A teacher's guide is available.

Who What When: Interactive Historical Timelines
http://www.sbrowning.com/whowhatwhen/index.php3
Access this database of people and events to create interactive graphic timelines.

The Wizard of Photography
http://www.pbs.org/wgbh/amex/Eastman
A companion to the film "The Wizard of Photography," this website includes transcripts, activities, a timeline of photography, a gallery of

Eastman Kodak advertisements, information on people and events in Eastman's life, and a teacher's guide.

Wright Brothers Aeroplane Company and Museum of Pioneer Aviation
http://www.wright-brothers.org
This virtual museum is organized into four sections: the history wing, the adventure wing, an information desk, and outreach programs. The museum guide and the search engine aid in navigating through the website.

Appendix C

Directory of Museums and Organizations

Contact addresses are listed for some of the museums and some of the organizations useful in the study of discoveries and inventions are listed below in alphabetical order. Many of them have websites for convenient access to information on their hours of operation and types of exhibits. Some of the museum websites have pictures of the exhibits, virtual tours, or online exhibits.

American Association for the Advancement of Science
1200 New York Avenue NW
Washington, DC 20005
Phone: 202-326-6400
E-mail: webmaster@aaas.org
http://www.aaas.org
Under the educational resources link is Science NetLinks that includes lessons, interactive activities for students, links to reviewed websites, and the benchmarks for science literacy.

American Computer Museum
Bridger Park Mall
2304 North 7th Avenue, Suite B
Bozeman, MT 59715
Phone: 406-582-1288
http://www.compuseum.org

Exhibits showing the evolution of the Information Age can be found in this museum. The museum honors living computer pioneers with the George R. Stibitz award. Past honorees and their contributions are listed on the website.

American Museum of Natural History
Central Park West at 79th Street
New York, NY 10024-5192
Phone: 212-769-5606
FAX: 212-769-5427
http://www.amnh.org/
This museum has a planetarium and a variety of exhibits including ones on mammals, birds, and fossils. The website has links to online explorations of some of the exhibits.

American Society of Inventors (ASI)
PO Box 58426
Philadelphia, PA 19102-8426
Phone: 215-546-6601
Email: info@asoi.org
http://www.americaninventor.org/
This organization was formed as a resource for inventors to provide them with information on patenting, developing, and marketing their inventions. Membership in the organization includes a subscription to *Inventor's Digest*. One section of the website contains current news articles of interest to inventors.

Astronomical Society of the Pacific
390 Ashton Avenue
San Francisco, CA 94112
Phone: 415-337-1100
FAX: 415-337-5205
http://www.astrosociety.org/
The purpose of this society is to serve as a liaison between astronomers and the public. It publishes an astronomy magazine, *Mercury*. The education link on the website provides resources for educators and students.

Carnegie Museum of Natural History
4400 Forbes Avenue
Pittsburgh, PA 15213

Phone: 412-633-3131
Email: cmnhweb@carnegiemuseums.org
http://www.carnegiemuseums.org/cmnh/
The museum features hands-on exhibits in the Discovery Room and has opportunities for distance learning that include video conferences with curators and scientists. The website has online activities and resources for educators.

Rachel Carson Homestead
613 Marion Avenue
Springdale, PA 15144
Phone: 412-274-5459
http://www.rachelcarson.org
This is the home of noted biologist, environmentalist, and writer Rachel Carson. She wrote about the interdependence of life on this planet and wanted everyone to develop an appreciation for the planet. Tours, environmental education classes, and a nature trail are available.

George Washington Carver Museum
1212 Old Montgomery Road
Tuskegee Institute, AL 36088
Mailing Address:
PO Drawer #10
Tuskegee Institute, AL 36087
George Washington Carver and Booker T. Washington are both given tributes at this historic site.

Center for the History of Physics
American Institute of Physics
One Physics Ellipse
College Park, MD 20740-3843
Phone: 301-209-3100
FAX: 301-209-0843
Email: aipinfo@aip.org
http://www.aip.org/history/ctrbro.htm
The Center mission is to preserve and to publicize the history of physics and allied fields. The web exhibits include information on Albert Einstein, Marie Curie, and the cyclotron, among other topics. There are resources for educators on the website.

Christopher Columbus Fellowship Foundation
110 Genesee Street, Suite 390
Auburn, New York 13021
Phone: 315-258-0090
FAX: 315-258-0093
Email: judithmscolumbus@cs.com
http://www.columbusfdn.org/
Established in 1992 to commemorate the 500th anniversary of the discovery of the Americas, this foundation provides encouragement and support to foster new discoveries that benefit humankind. The foundation established the National Gallery for America's Young Inventors, which each year inducts up to six young inventors in grades kindergarten through twelfth grade.

Computer Museum of America
640 C Street
San Diego, CA 92101
Phone: 619-235-8222
Email: dweil@computer-museum.org
http://www.computer-museum.org
The online exhibit hall contains pictures of some of the computers currently exhibited in the museum. This museum has a Hall of Fame, and information on the inductees can be found on the website.

Cousteau Society
710 Settler Landing Road
Hampton, VA 23669
Phone: 800-441-4395/757-722-9300
FAX: 757-722-8185
Email: cousteau@cousteausociety.org
The purpose of the Cousteau Society is to foster understanding and appreciation for the planet's water systems to assure they will be protected. A kid's corner and information about Captain Jacques-Yves Cousteau can be found on the site, which is available in English or French.

Charles Darwin Foundation for the Galapagos Islands
Av. 6 de Diciembre N 36-109 y Pasaje California
Post Box 17-01-3891
Quito, Ecuador

Charles Darwin Research Station
Puerto Ayora, Santa Cruz
Galapagos Islands, Ecuador
Phone: 593-5526-147/148, 593-5527-013/014
http://www.darwinfoundation.org/
The Charles Darwin Foundation is dedicated to the conservation of the
Galapagos ecosystems and operates the Charles Darwin Research Station. Information about ongoing research and a Kids' Korner can be
found on the website. The information is available in English, Spanish,
French, and Dutch.

George Eastman House International Museum of Photography and
Film
900 East Avenue
Rochester, NY 14607
Phone: 716-271-3361
http://Eastman.org
Eastman's house and a museum are located on this twelve and one-half
acre site. There are exhibits, galleries, theaters, an archives building/research center, and an education center.

Thomas A. Edison Birthplace Museum
9 Edison Drive
Milan, OH 44846
Phone: 419-499-2135
http://www.tomedison.org/
Period furniture, Edison family memorabilia, and an exhibit of some of
Edison's inventions are available for viewing in this museum.

Thomas Edison Butchertown House
729-31 East Washington Street
Louisville, KY 40202
Phone: 502-585-5247
This is the cottage where Edison stayed while a telegrapher in Louisville. Exhibits include Edison memorabilia and inventions.

Thomas A. Edison Memorial Tower and Menlo Park Museum
Route 27
Edison, NJ 08817
Phone: 201-549-3299
http://www.edisonnj.org/menlopark/museum.asp

There is a 131-foot tower commemorating Edison's Menlo Park laboratory. The museum houses some of Edison's inventions and products made by the Thomas A. Edison Company. On the website there are audio clips of vintage recordings.

Edison National Historic Site
Main Street and Lakeside Avenue
West Orange, NJ 07052
Phone: 201-736-5050
http://www.nps.gov/edis/home.htm
Operated by the National Park Service, this historic site contains Edison's laboratory and is the repository for many of his papers and records.

Environmental Protection Agency
401 M Street SW
Washington, DC 20460
http://www.epa.gov/
On the website under the link for educational resources there is information for kids, for students, and for teachers. The website can be viewed in either English or Spanish.

Exploratorium: The Museum of Science, Art, and Human Perception
3601 Lyon Street
San Francisco, CA 94123
Phone: 415-561-0360
TTY: 415-353-0400
http://www.exploratorium.edu/
Noted for its interactive science demonstrations, this museum also houses a theater and the Tactile Dome for exploring with the sense of touch.

Fermi National Accelerator Laboratory and the Leon M. Lederman Science Education Center
PO Box 500
Batavia, IL 60510-0500
Phone: 630-840-3351
FAX: 630-840-8780
http://www.fnal.gov/directorate/public_affairs
This laboratory conducts research in high-energy physics. Basic science history and atomic news, such as the discovery of new particles,

can be found on the website. The site includes a virtual tour of the lab and resources for teachers. Its Science Education Center has hands-on exhibits for kindergarten through twelfth grade students.

Field Museum of Natural History
1400 South Lake Shore Drive
Chicago, IL 60605
Phone: 312-922-9410
http://www.fmnh.org/
At this museum visitors can see Sue, the world's largest and most complete Tyrannosaurus rex fossil. Wilderness exhibits and an Ancient Egypt exhibit include hands-on explorations. The website has online resources for students and teachers.

Henry Ford Estate: A National Historic Landmark
University of Michigan at Dearborn
4901 Evergreen Road
Dearborn, MI 48128-1491
Phone: 313-593-5590
http://www.umd.umich.edu/fairlane
Henry Ford and his wife lived in this home for over thirty years. Visitors can tour the home and the gardens. The website for this museum includes lesson plans, student activities, research archives, and oral histories.

The Henry Ford Museum and Greenfield Village
PO Box 1970
20900 Oakwood Boulevard
Dearborn, MI 48121-1620
Phone: 800-835-5237/313-271-1620
http://www.hfmgv.org
This museum is a celebration of industry and technology and includes reconstructions of a variety of historic buildings including Menlo Park, the Wright brothers' bicycle shop, and a courthouse where Abraham Lincoln practiced law.

The Franklin Institute
222 North 20th Street
Philadelphia, PA 19103
Phone: 215-448-1200
http://www.fi.edu

This hands-on museum is a testimony to the wisdom and inventiveness of Benjamin Franklin. In addition to numerous interesting exhibits, the museum houses a theater.

Intellectual Property Owners
1255 Twenty-third Street NW, Suite 850
Washington, DC 20037
Phone: 202-466-2893
FAX: 202-466-2893
Email: info@ipo.org
http://www.ipo.org/
This nonprofit organization's members hold patents, trademarks, and copyrights. The Spirit of America Ingenuity Award and the National Inventor of the Year Award are contests sponsored by this organization.

International Federation of Inventors' Associations (IFIA)
IFIA Secretariat
PO Box 299
1211 Geneva 12
Switzerland
FAX: (41 22) 789 3076
Email: invention-ifia@bluewin.ch
http://www.invention-ifia.ch/index.html
This international organization works to promote cooperation between inventor associations worldwide.

The Jenner Museum
Church Lane, Berkeley
Gloucestershire, GL13 9BH
England
Phone: (44 0) 1453 810631
FAX: (44 0) 1453 811690
http://www.jennermuseum.com
This museum is dedicated to Edward Jenner and the smallpox vaccine. The website contains information on his life, his research, immunology, and some games.

Jet Propulsion Laboratory
4800 Oak Grove Drive
Pasadena, CA 91109
Phone: 818-354-4321

http://www.jpl.nasa.gov/
This laboratory houses NASA's planetary exploration program. The Spacecraft Museum, the Space Flight Operations Facility, and the Spacecraft Assembly Faculty are open to visitors. The website includes resources for students and educators.

Lyndon B. Johnson Space Center
2101 NASA Road 1
Houston, TX 77058
Phone: 281-483-0123/281-483-8693
http://www.jsc.nasa.gov/
The Space Center sponsors programs for students and educators. The website includes press releases, photographs, and other information about the space program.

Lawrence Hall of Science (LHS)
University of California, Berkeley
Centennial Drive
Berkeley, CA 94720-5200
Phone: 510-642-5132
Email: lhsinfo@uclink.berkeley.edu
http://www.lhs.berkeley.edu/
LHS creates model programs that focus on the teaching and the learning of science and mathematics for preschool through twelfth grade. Information on these materials and on current exhibits is available on the website.

Jerome and Dorothy Lemelson Center for the Study of Invention and Innovation
National Museum of American History, Suite 1016
Smithsonian Institution
Washington, DC 20560-0604
Phone: 202-357-1593
Email: lemcen@nmah.si.edu
http://www.si.edu/lemelson/
The center was established to foster inventive creativity in children and to make them aware of the importance of inventions and innovations in American history. The website has resources for educators to use in their classrooms, a section on Thomas Edison, and virtual exhibits.

Lawrence Livermore National Laboratory Discovery Center
7000 East Avenue
Livermore, CA 94550-9234
Phone: 925-422-5815
http://www.llnl.gov/
The Discovery Center exhibits highlight the research and history of the
lab in areas including defense, homeland security, biomedicine, environmental science, and new energy sources.

Massachusetts Institute of Technology Museum
265 Massachusetts Avenue
Cambridge, MA 02139
Phone: 617-253-4444
FAX: 617-253-8994
http://web.mit.edu/museum/
The museum showcases the research being done at MIT in science and
technology. The website contains links to information about current
exhibits and educational programs.

Cyrus McCormick Farm and Workshop
PO Box 100
128 McCormick Farm Circle
Steele's Tavern, VA 24476
Phone: 540-377-2255
FAX: 540-377-5850
Email: dafiske@vt.edu
http://www.vaes.vt.edu/steeles/mccormick/mccormick.html
Visitors can tour the workshop where McCormick invented the me-
chanical reaper. The website had a video clip of the McCormick Grist
Mill Water Wheel, a biography of McCormick, and a history of grain
harvesting.

Maria Mitchell Association
4 Vestal Street
Nantucket, MA 02554
Phone: 508-228-2896/508-228-9198
FAX: 508-228-1031
http://www.mmo.org
This association honors Maria Mitchell, the first professional woman
astronomer. The association strives to enhance understanding of the
universe through research and education. The Maria Mitchell house is

open to visitors and information about visiting can be obtained from the association website.

Mount Wilson Observatory
PO Box 60947
Pasadena, CA 91116
Phone: 818-793-3100
FAX: 818-793-4570
http://www.mtwilson.edu
Visit this observatory in person or virtually via the website. The focus of this observatory is on solar research.

Museum of Broadcast Communications
78 East Washington Street
Chicago, IL 60602-4801
Phone: 312-629-6000
http://www.museum.tv/
The museum features interactive exhibits and archives of television and radio programs. On the website educators find resources for use in classrooms and libraries such as downloadable lesson plans and streaming video.

Museum of Paleontology, University of California–Berkeley
1101 Valley Life Sciences Building #4780
Berkeley, CA 94720-4780
Phone: 510-642-1821
FAX: 510-642-1822
http://www.ucmp.berkeley.edu
Information about current museum research and exhibits is available on the website. The online exhibits contain information on phylogeny, geologic time, and evolutionary thought.

Museum of Science, Science Park
Boston, MA 02114
Phone: 617-723-2500
TTY: 617-589-0417
Email: information@mos.org
http://www.tcm.org
The focus of the museum is on science and technology. Information on current exhibits and online exhibits are available on the website. Online exhibits include the Scanning Electron Microscope; Theater of Elec-

tricity; Leonardo da Vinci: Scientist, Inventor, Artist; and Design Your Own Robot.

Museum of Science and Industry
57th Street and Lake Shore Drive
Chicago, IL 60637-2093
Phone: 800-468-6674/773-684-1414
TDD: 773-684-3323
http://www.msichicago.org/
Inspiring the inventive genius in everyone is the mission of this museum. There are a number of permanent exhibits including ones on genetics, trains, and submarines. Online versions of these exhibits can be accessed from the website.

Museum of Television and Radio
New York Location:
25 West 52nd Street
New York, NY 10019
Phone: 212-621-6600
Los Angeles Location:
465 North Beverly Drive
Beverly Hills, CA 90210
Phone: 310-786-1025
http://www.mtr.org
The museum offers daily screenings and radio presentations, exhibits, seminars, and a listening room. The website contains information for both locations.

National Air and Space Museum, Smithsonian Institution
Sixth Street and Independence Avenue, SW
Washington, DC 20560
Phone: 202-357-1686
http://www.nasm.si.edu
This museum houses exhibits marking the history of aviation and space flight as well as a theater and a planetarium. The website has links to photographs and information on current exhibits, former exhibits, online exhibits, and museum collections.

NASA Goddard Space Flight Center
Code 130, Public Affairs Office
Greenbelt, MD 20771

Phone: 301-286-8955
http://www.gsfc.nasa.gov/
Scientists and engineers at this center are exploring the Earth from space. The visitor center offers presentations, tours, and resources for educators. The website includes information on current NASA events, space science missions, the results from studies of the Earth, and a brief biography of Robert Goddard.

National Aviation Hall of Fame
1100 Spaatz Street
Wright Patterson Air Force Base, Ohio 45433
Mailing Address:
PO Box 31096
Dayton, Ohio 45437
Phone: 937-256-0944
FAX: 937-256-8536
http://www.nationalaviation.org/
Onsite and online exhibits include Early Years, World War I, the Golden Age, World War II, the Jet Age, and the Space Age. The website includes resources for educators and students.

National Inventors Hall of Fame
221 South Broadway
Akron, OH 44308-1505
Phone: 330-762-4463
FAX: 330-762-6313
Email: museum@invent.org
http://www.invent.org/
The museum contains exhibits on inventors and their inventions. The Hall of Fame conducts workshops and sponsors an inventors' camp for children.

National Museum of American History, Smithsonian Institution
14th Street and Constitution Avenue, NW
Washington, DC 20560
Phone: 202-357-2700
FAX: 202-633-9338
TTY: 202-357-1729
http://americanhistory.si.edu/
The museum houses three floors of exhibits, a hands-on history room, and a hands-on science room. There are virtual exhibits on the website.

National Museum of Natural History, Smithsonian Institution
Tenth Street and Constitution Avenue, NW
Washington, DC 20560
Phone: 202-357-2700
Email: info@info.si.edu
http://www.mnh.si.edu
Visit the live insect museum and attend performances and lectures in the Baird Auditorium. The museum also houses the Discovery Center and a Naturalist Center.

National Museum of Science and Industry
Exhibition Road
South Kensington
London SW7 2DD
United Kingdom
Phone: (44 171) 938 8008/8080
Disabled Persons Inquiry Line: (44 171) 938 9788
http://www.nmsi.ac.uk
This family of museums includes the Science Museum, Creative Planet, the National Railway Museum, and the National Museum of Photography, Film, and Television. Information on each of these museums can be downloaded from the website. The museums are in the process of digitizing their collections.

National Science Teachers Association (NSTA)
1840 Wilson Boulevard
Arlington, VA 22201-3000
Phone: 703-243-7100
http://www.nsta.org/index.html
NSTA provides resources to foster the teaching and learning of science. This organization publishes four journals for its members: *Science and Children*, *Science Scope*, *The Science Teacher*, and the *Journal of College Science Teaching*. The website contains resources for teachers and links to other sites with additional resources.

National Space Society
600 Pennsylvania Avenue, SE, Suite 201
Washington, DC 20003
Phone: 202-543-1900
FAX: 202-546-4189
Email: nsshq@nss.org

http://www.nss.org
Werner von Braun founded this organization in 1974. Its purpose is to work toward the day when people will live and work in space. The website includes resources for educators and students.

National Toy Hall of Fame
A.C. Gilbert's Discovery Village
116 Marion Street, NE
Salem, OR 97301-3437
Phone: 800-316-3485
FAX: 503-316-3485
http://www.acgilbert.org
Displayed alongside the toys inducted into the Hall of Fame are a select group of toys invented by children. They submit a picture and a written description of the toy they invented in order to have it considered for the honor.

National Wildlife Federation
11100 Wildlife Center Drive
Reston, VA 20190-5362
Phone: 703-438-6000
http://www.nwf.org/
The National Wildlife Federation works to ensure the conservation of wildlife and their habitats. It provides school and community programs to promote environmental education. The Federation publishes magazines including *National Wildlife*, *Ranger Rick*, *Your Big Backyard*, and *Wild Animal Baby*. It also produces films to highlight the importance of conservation.

Nobel Museum
Box 2245
103 16 Stockholm
Sweden
Phone: (46 0) 8-23 25 06
FAX: (46 0) 8-23 25 07
http://www.nobel.se/nobel/nobelmuseum/index.html
Visitors learn about twentieth-century discoveries, Alfred Nobel, the Nobel Prize, and the Laureates.

Sierra Club
National Headquarters
85 Second Street, 2nd Floor
San Francisco, CA 94105
Phone: 415-977-5500
FAX: 415-977-5799
Email: information@sierraclub.org
www.sierraclub.org
One of the founding members of this environmental organization was
John Muir. The website includes updates on environmental issues and
information on activities in each state. Membership in this organization
includes a subscription to *Sierra* magazine. The website is available in
English and Spanish.

The Skyscraper Museum
55 Broad Street, #13F
New York, NY 10004
Phone: 212-968-1961
http://www.skyscraper.org
The purpose of the museum is to study high-rise buildings of the past,
the present, and the future. The website features an interactive timeline
of skyscrapers and an archive of previous exhibits.

Society for Amateur Scientists
5600 Post Road, Suite 114-341
East Greenwich, RI 02818
Phone: 401-823-7800
FAX: 401-823-6800
Email: info@sas.org
http://www.sas.org/
This group provides support to amateur scientists and their explora-
tions.

Space Center Houston
2101 NASA Road One
Houston, TX 77058
Phone: 281-244-2100
FAX: 281-283-7724
http://www.spacecenter.org

Tours of the center include exhibits about humans in space and the Apollo missions. The website includes resources for educators and information about programs for students.

Space Telescope Science Institute
3700 San Martin Drive
Johns Hopkins University Homewood Campus
Baltimore, MD 21218 USA
Phone: 410-338-4700
http://www.stsci.edu
Operated by the Association of Universities for Research in Astronomy, Inc. (AURA), the Space Telescope Science Institute (STScI) has established partnerships with NASA. It is responsible for the scientific operation of the Hubble Space Telescope. The website houses searchable astronomical data archives.

Spaceport USA
Kennedy Space Center, FL 32899
Phone: 321-867-5000
http://www.ksc.nasa.gov/
Spaceport includes a visitor center with a theater, an outdoor exhibit of rockets, and indoor exhibits about human space flight.

Springfield Armory National Historic Site
One Armory Square
Springfield, MA 01105-1299
Phone: 413-734-8551
FAX: 413-747-8062
http://www.nps.gov/spar/
The U.S. Army conducted research and developed small firearms at this site until 1967. Inventions developed at this site impacted not only weaponry but many consumer products as well.

The Tech Museum of Innovation
201 South Market Street
San Jose, CA 95113
Phone: 408-294-8324
http://www.thetech.org/
The museum houses a number of interactive exhibits showcasing high-tech technology. The website features online exhibits including The Spirit of American Innovation: The National Medal of Technology.

This exhibit has information on the winners of the medal. Online teacher resources include lesson plans for Design Challenge, which involves students in designing a solution to an authentic real-world problem.

United States National Library of Medicine
8600 Rockville Pike
Bethesda, MD 20894
Phone: 888-346-3656/301-594-5983
FAX: 301-402-1384
http://www.nlm.nih.gov/nlmhome.html
This is the world's largest biomedical library and many of its resources from medical history to biotechnology are available on the website. Medline Plus Health Information can be accessed from the website.

United States Patent and Trademark Office Museum
2121 Crystal Drive, Suite 0100
Arlington, VA 22202
Phone: 800-786-9199/703-306-3457
FAX: 703-308-5258
http://www.uspto.gov/web/offices/ac/ahrpa/opa/museum/
The museum was established to help people understand the patent and trademark system and to understand the role of intellectual property protection. Information on exhibits and arranging tours can be found on the website.

White Sands National Monument
PO Box 1086
Holloman Air Force Base, NM 88330
Phone: 505-479-6134
http://www.nps.gov/whsa/
This is the site where the first atomic bomb was exploded. Visitors can tour the building where the bomb was assembled. The geology and natural beauty of this site also attract visitors.

Women Inventors Project
1 Greensboro Drive, Suite 302
Etobicoke, Ontario
Canada M9W 1CB
Phone: 877-863-2471/905-731-0328
FAX: 905-731-9691

email: womenip@interlog.com
http://www.womenip.com/
This organization promotes women inventors and entrepreneurs in Canada, the United States, and internationally. It also develops materials, publications, and programs to heighten public awareness of female inventors and entrepreneurs. The website includes profiles of women inventors.

World Health Organization (WHO)
Avenue Appia 20
1211 Geneva 27
Switzerland
Phone: (41 22) 791 21 11
FAX: (41 22) 791 3111
http://www.who.int/en/
Established by the United Nations in 1948 the World Health Organization's mission is to assure the highest possible level of health for all people. The website contains extensive information on medical and health issues.

World Future Society
7910 Woodmont Avenue, Suite 450
Bethesda, MD 20814
Phone: 800-989-8274/301-656-8274
FAX: 301-951-0394
Email: info@wfs.org
http://www.wfs.org
This is a nonprofit, educational and scientific organization that serves as a clearinghouse for ideas about the future. Its members are interested in examining the social and technological developments that will shape the future.

The Wright Brothers National Memorial
c/o Cape Hatteras National Seashore
Route 1, Box 675
Manteo, NC 27954
Phone: 919-441-7430
FAX: 919-441-7730
http://www.nps.gov/wrbr

A monument on Kill Devil Hill commemorates the Wright brothers' flights. This memorial also includes a museum and reconstructed camp buildings.

Youth Science Foundation Canada
481 University Avenue, Suite 703
Toronto, Ontario M5G 2E9
Phone: 866-341-0040/416-341-0040
FAX: 416-341-0041
Email: info@ysf.ca
http://www.ysf.ca/
Motivating young Canadians to take an interest in science and technology is the mission of this organization. One of the outreach programs is the Intel International Science and Engineering Fair.

Appendix D

Booktalks, Classroom Activities, and Invention Contests

The booktalks, classroom activities, and invention contests in this appendix provide educators with ideas for using the materials annotated in this book. Teachers and librarians are encouraged to adapt the booktalks and activities to their students' needs and abilities. Chapter 18 Invention to Market and Appendix A Professional Resources for Educators include books and articles that are useful resources for educators and students as they create their own inventions.

Booktalks

Giving booktalks is one of the best ways to share books with students and to motivate them to read. While most booktalks are prepared for and shared with young adults, there are many books for younger children that can also be "booktalked." The key to a good booktalk is knowing how much to tell and how much to leave out. Enough of the story must be told to whet the appetite of the reader and enough left untold to make the reader want to know more. Below are several different booktalks that may help guide you as you prepare your own.

Compestine, Ying Chang. *The Story of Paper*. New York: Holiday House, 2003. 32 pp. Gr. K–2.

The next day on the way to school, Kuai told his brothers, "I have an idea for how to make something to write on."

"What? How?" asked Ting and Pan with great interest.

"Did you notice how smooth the surfaces of the rice cakes were yesterday? They were almost as smooth as the surface of silk," said Kuai.

"But you can't fold a rice cake. It's thick and sticky," said Ting.

"I don't want to write on a rice cake. I want to eat it!" Pan smacked his lips.

"What if we use something other than rice?" asked Kuai. He told his brothers his plan.

Kuai, Ting, and Pan were three Chinese brothers who were always being scolded at school because they were always so curious about so many things that they never paid proper attention to the teacher. Because paper had not yet been invented, the teacher had to let the parents know of their behavior by writing messages on their hands. Since most people wrote on silk, only the very wealthy had anything to write on. One day, when Mother was making rice cakes, Kuai had an idea that he and his brothers could try. Read *The Story of Paper* to see if his idea worked.

Jones, Charlotte Foltz. *Mistakes That Worked*. New York: Doubleday, 1991. Gr. 6–12.

The dictionary's first definition of the word "fudge" is "nonsense or foolishness." And that's how our favorite chocolate candy got its name. A story says that in the 1890s, a candy-maker in Philadelphia, Pennsylvania, was supervising his employees as they made caramels. Someone made a mistake and instead of producing a chewy candy, the batch turned into a finely crystallized, nonchewy substance. "Fudge!" the candy-maker swore. And with that exclamation to describe the mistake, fudge was born.

Did you know that potato chips were first made by an angry chef when a customer complained that his fried potatoes were not cut thin enough or fried long enough? Did you know that donuts with holes in the middle are reported to have been invented by a sea captain who forced the fried cake he was eating down on the spokes of the ships wheel? Would you like to know the origin of trouser cuffs? These are just a few of the things that were invented "by mistake." Learn where aspirin, x-rays, and Silly Putty came from. Read *Mistakes That Worked* by Charlotte Jones.

Tambini, Michael. *Future.* New York: Dorling Kindersley, 2000. Gr. 6–12.

Arthur C. Clarke has written many science fiction novels about space exploration. He also predicted the use of satellites for global communications.

In 1953, Francis Crick and James Watson discovered the molecular structure of DNA. The genetic code for all life is contained within this molecule. Our ability to understand and manipulate it will be central to the 21st century.

Virtual reality is already being used for entertainment, as well as for medicine and design. In the future, virtual reality will become as familiar to us as movies.

This title, *Future,* provides a look ahead to the technological, environmental, and biological developments of the twenty-first century. Buildings, transportation, robots, genetic engineering, cyber-body parts, and environmental-friendly activities are all described and discussed as prophetic illusions of the 2000s. The illustrations are mostly brightly colored photographs, which will whet the appetites of today's readers. Read *Future* and see if your appetite is whetted.

Tucker, Tom. *Brainstorm! : The Stories of Twenty American Kid Inventors.* New York: Farrar, Straus and Giroux, 1995. Gr. 6–12.

On the first Friday in January 1991, Vanessa Hess sat in her seventh-grade classroom at Stonybrook Junior High and heard her teacher, Mrs. Maurine Marchani, announce that each kid had to do a new project—they had to invent something. "There are only two ways you can avoid this," Mrs. Marchani said. "You can die, or you can move."

There it was. The boy held the concoction in his hand: a glass containing flavored soda water left outside overnight. It had frozen. All over San Francisco, a cold snap had worked its magic and even the surface of Stow Lake in Golden Gate Park had turned solid. His mixing stick was frozen into the mixture, too, and now protruded like a handle. But the warmth from the palm of his hand was enough: the icy mass slipped free of the glass and his fingers clutched the wooden stick. The year was 1905. At age eleven, Frank W. Epperson was holding the world's first Popsicle.

These interesting stories of young people and their inventions are told in an entertaining fashion. In addition to twenty surprising stories of gumption, hard work, and ingenuity triumphing over false starts and long odds, *Brainstorm!* contains concise background information about the American patent system and step-by-step advice to young inventors for turning their own great ideas into great inventions of the future.

Cross, Gillian. *New World.* New York: Holiday House, 1994. 171 pp. Gr. 7–12.

> Miriam looked down at the keyboard on top of the game tower. "I tell it my name first. Right?" Hesketh nodded gravely, and she typed in the letters, M-I-R-I-A-M. Then she took the case and flicked the catches open. Miriam's skin tingled and she stared down into the case. All the kit was there, each piece in its own slot. The single glove. The belt. The Game Helmet—like heavy goggles, with earpieces and interwoven straps and three clumsy bracelets. She took the bracelets out first. As she clicked them into place—one round her left wrist and the other two round her ankles—she could almost hear Christine Riley chuckle. *Virtual jewelry they call this. I can't say I care for Hesketh's taste, but they're not meant to look pretty. They help to tell the computer where you are.*

Imagine that you are a fourteen-year-old student who has been asked by a video game company to test one of their new virtual reality games before they market it to the public. You are told that you and a fourteen-year-old boy have been selected for this honor "at random." You only have to give an hour, three days a week, for several weeks AND you will be PAID for your time! Miriam thinks she is the luckiest girl in the world. And the only restriction is that she is not allowed to tell ANYONE what she is doing or anything about the game. The game is terrific fun and she loves the challenge. But suddenly the most frightening thing that can happen to Miriam DOES. Her terrifying recurring nightmare that NO ONE but she and her father know about happens in the game! Miriam has given her word that she will tell no one about the game, but what is happening to her? Find out by reading *New World* by Gillian Cross.

Classroom Activities

In this section are ideas for classroom activities using some of the materials annotated in this book. These activities include creating composite inventions, making a list of the greatest medical discoveries, the Inventive Thinking Curriculum Project, and constructing timelines. These activities provide students opportunities to work in groups and involve a hands-on approach to learning about discoveries and inventions. The activities can be adapted for different grade levels and different abilities.

Composite Inventions

Not all inventions are brand new ideas, some inventions make improvements to existing ones. For example, Robert Fulton did not invent the steamboat, he made improvements to the design to make it financially viable. Some inventions are not used only as their inventors intended, instead other uses are found for them. For example, microwaves were first used in Britain's radar system to detect Nazi warplanes. Later, it was discovered that microwaves could cook foods and the microwave radar range was invented. This activity challenges students to think creatively about ways to combine two items and develop a composite invention. This may involve using one invention to improve another invention, or combining the two inventions to do something that neither of the inventions presently does.

Books

Perry, Andrea. *Here's What You Do When You Can't Find Your Shoe: Ingenious Inventions for Pesky Problems*. Gr. K–5.

Harper, Charise Mericle. *Imaginative Inventions: The Who, What, Where, When and Why of Roller Skates, Potato Chips, Marbles, and Pie*. Gr. 1–4.

Jones, Charlotte Foltz. *Accidents May Happen: Fifty Inventions Discovered by Mistake*. Gr. 3–8.

Wilsdon, Christina. *Everyday Things: An A–Z Guide*. Gr. 4–8.

Flatow, Ira. *They All Laughed ... From Light Bulbs to Lasers: The Fascinating Stories Behind the Great Inventions That Have Changed Our Lives*. Gr. 7–12.

Wolfe, Maynard Frank, and Rube Goldberg. *Rube Goldberg: Inventions!* Gr. 7–12.

Gratzer, Walter. *Eurekas and Euphoria: The Oxford Book of Scientific Anecdotes.* Gr. 9–12.

Lindsay, David. *House of Invention: The Secret Life of Everyday Products.* Gr. 10–12.

Materials

1. Mail order catalogs
2. Paper and pencils

Activity

Sharing some of the books noted above with the students may help them begin to think of unique ways to use items. Then, students randomly select two pages from a large mail order catalog. They select one item from each page and combine the two items into a unique and useful invention. Students draw their composite invention and write a description telling how it is used. Younger students may need to see this activity modeled in a whole class setting before working with a partner to devise their own invention. This activity is adapted from *The Art and Science of Invention* (Kivenson, 1982), which is out of print.

Greatest Medical Discoveries

Creating a list of five or ten of the greatest medical discoveries requires students to research medical discoveries and critically think about the impact of the discoveries on people's health and well-being. Once they have made their choices they need to be able to defend them. One of the books listed below contains the ten greatest medical discoveries, but students should think critically about medical discoveries and create their own lists. Variations on this could include determining the most important household inventions, the most useful byproducts developed from space exploration inventions, or the most important communication inventions.

Books

Darling, David. *The Health Revolution: Surgery and Medicine in the Twenty-First Century.* Gr. 4–8.

Parker, Steve. *Medical Advances.* Gr. 4–8.

Donnellan, William L. *The Miracle of Immunity.* Gr. 5–8.

Himrich, Brenda L., and Stew Thornley. *Electrifying Medicine: How Electricity Sparked a Medical Revolution.* Gr. 5–8.

Miller, Brandon Marie. *Just What the Doctor Ordered: The History of American Medicine.* Gr. 5–8.

Yount, Lisa. *History of Medicine.* Gr. 6–9.

Hyde, Margaret O., and John F. Setaro. *Medicine's Brave New World.* Gr. 6–12.

Nardo, Don. *Vaccines.* Gr. 7–10.

Friedman, Meyer, and Gerald W. Friedland. *Medicine's Ten Greatest Discoveries.* Gr. 9–12.

Smolan, Rick, and Phillip Moffitt, eds. *Medicine's Great Journey: One Hundred Years of Healing.* Gr. 9–12.

Websites

Center for Bioethics
http://bioethics.net/hsbioethics/
Doctor Over Time
http://www.pbs.org/wgbh/aso/tryit/doctor/
ProMedicine through Time
http://www.bbc.co.uk/education/medicine/nonint/home.shtml
National Center for Biotechnology Information
http://www.ncbi.nlm.nih.gov/
National Human Genome Research Institute
http://www.genome.gov/

Articles

Banchereau, Jacques. "The Long Arm of the Immune System." *Scientific American* 287, no. 5 (Nov. 2002): 52–59.

"History's Top Invention." *Science World* 59, no. 12 (Mar. 28, 2003): 7.

Langreth, Robert. "The Doctor Is In." *Forbes* 172, no. 4 (Sept. 1, 2003): 88–89.

"Prototype." *Technology Review* 106, no. 4 (May 2003): 14–16.

Reed, J.D. "Virus Vanquisher." *Smithsonian* 32, no. 11 (Feb. 2002): 28–29.

Sheppard, Robert. "Medicine in 2020." *Maclean's* 113, no. 3 (Jan. 10, 2000): 40–43.

"Speed Reading the Book of Life." *Economist* 367, no. 8329 (June 21, 2003): 13.

Trachtman, Paul. "Hero for Our Time: Louis Pasteur and Anthrax." *Smithsonian* 32, no. 10 (Jan. 2002): 34–41.

Wilson, Jim. "No More Needles." *Popular Mechanics* 180, no. 6 (June 2003): 42–44.

Materials

1. Books, websites, and articles
2. Paper and pencil
3. *PowerPoint* presentation software from Microsoft Corporation

Activity

Begin by having students brainstorm a list of medical discoveries. Using the list, have them create a survey to give to their classmates and family members to get their opinions on the greatest medical discoveries. Then, have the students work in small groups to research medical discoveries and have each group create their own list of the greatest medical discoveries. They should be able to justify why each discovery was included on their list. Provide the groups with opportunities to share their lists with the whole class. One way to share their lists is by creating *PowerPoint* presentations of the lists with brief justifications for their selections.

Inventive Thinking Curriculum Project

This project is an outreach program of the United States Patent and Trademark Office and is for use as part of a thinking skills program. It requires children to apply their critical thinking skills, their creative thinking skills, and their problem-solving skills as they create an invention or an innovation. A packet of materials can be downloaded from the website listed below. Included in the packet are an introduction to thinking skills models, nine activities that culminate in a Young Inventor's Day, a suggestion for parent involvement, an enrichment copyright primer and an appendix that contains copies of selected patents by noted inventors including among others Thomas Alva Edison, Granville T. Woods, and An Wang. This project can be adapted for different grade levels.

Website

Inventive Thinking Curriculum Project
http://www.uspto.gov/go/opa/projxl/invthink/invthink.htm

Materials

1. Activities from website
2. Blackline masters from website

Activity

Complete the first three activities to help students practice their creative thinking skills. Activities four through nine move them through the invention process. Activity ten contains suggestions for parental involvement and activity eleven has suggestions for hosting a Young Inventor's Day.

Timelines

As students read biographies of discoverers and inventors, they learn that the information in the biographies may be conflicting and that some biographies focus more on the person's life and others focus more on the person's work. Some may focus on the person's youth or focus on the person as an adult. Reading more than one biography enables students to fully appreciate the contributions of the person. As students learn about discoverers and inventors, it is important that they be able to place them in the context of the times in which they lived. One way to do this is to have the students construct timelines of the life of the discoverer or inventor including information about the world during that person's lifetime.

Books

Ford, Carin T. *Alexander Graham Bell: Inventor of the Telephone*. Gr. 2–4.
Cefrey, Holly. *The Inventions of Alexander Graham Bell: The Telephone*. Gr. 2–6.
Fisher, Leonard Everett. *Alexander Graham Bell*. Gr. 3–5.
Pollard, Michael. *Alexander Graham Bell: Father of Modern Communication*. Gr. 6–10.
St. George, Judith. *Dear Mr. Bell . . . Your Friend, Helen Keller*. Gr. 7–12.

McLoone, Margo. *George Washington Carver*. Gr. 1–3.
Adler, David A. *A Picture Book of George Washington Carver*. Gr. 1–4.
Kramer, Barbara. *George Washington Carver: Scientist and Inventor*. Gr. 3–5.

Fisher, Leonard Everett. *Marie Curie*. Gr. 2–5.
Parker, Steve. *Marie Curie and Radium*. Gr. 3–8.
Pflaum, Rosalynd. *Marie Curie and Her Daughter Irene*. Gr. 5–8.
Poynter, Margaret. *Marie Curie: Discoverer of Radium*. Gr. 5–12.
Labouisse, Eve Curie. *Madame Curie: A Biography*. Gr. 9–12.

Dolan, Ellen M. *Thomas Alva Edison: Inventor*. Gr. 5–12.
Tagliaferro, Linda. *Thomas Edison: Inventor in the Age of Electricity*. Gr. 6–8.
Israel, Paul. *Edison: A Life of Invention*. Gr. 10–12.

Fritz, Jean. *What's the Big Idea, Ben Franklin?* Gr. 1–3.
Giblin, James Cross. *The Amazing Life of Benjamin Franklin*. Gr. 3–5.
Brands, H. W. *The First American: The Life and Times of Benjamin Franklin*. Gr. 10–12.

Videos

Alexander Graham Bell: Voice of Invention from A and E Television Networks.
Marie Curie More Than Meets the Eye from Devine Entertainment.
Edison: The Wizard of Light from Devine Entertainment.
Ben Franklin: Citizen of the World from A and E Television Networks.

Websites

Biography.com
http://www.biography.com
The Black Inventor Online Museum
http://www.blackinventor.com
EnchantedLearning.com: Zoom Inventors and Inventions
http://www.enchantedlearning.com/inventors/indexa.shtml
Kids Cafe: Patent Café's Space for Young Inventors
http://kids.patentcafe.com/index.asp
People and Discoveries
http://www.pbs.org/wgbh/aso/databank/
Who What When: Interactive Historical Timelines
http://www.sbrowning.com/whowhatwhen/index.php3

Materials

1. Books, videos, and websites
2. Paper, pencil, and markers for creating a timeline
3. *Timeliner* software from Tom Snyder Productions

Activity

Using a variety of resources, have the students gather information on a discoverer or an inventor. Then, have the students gather information on world events during the person's lifetime. One place to find this information is *Who What When: Interactive Historical Timelines* available at http://www.sbrowning.com/whowhatwhen/index.php3. Using a chart such as the one below for gathering information may be helpful to the students. (See figure D.1.) They can then enter the information in *Timeliner* or create a timeline by taping sheets of paper together. Encourage the students to add graphics to their timeline to illustrate key points or events in the person's life. Younger students may create a visual timeline using only graphics. The program contains graphics and students can also import their own graphics into the timelines.

Timeline Chart

Year	Events in the life of the Inventor/Discoverer	World Events
	Birth	
	Death	

Figure D.1. Students can use a timeline chart to record and organize the information they gather.

Invention Contests

There are several invention contests sponsored by different organizations that are open to students. This section contains brief descriptions of some of the contests and includes contact information for learning more about the contests. The previous section, Classroom Activities, contains information on the "Inventive Thinking Curriculum Project" that has resources for leading students through the invention process. Chapter 18 Invention to Market contains books about young inventors and their inventions as well as books with information on the invention process. Appendix A Professional Resources for Educators has references to articles about teaching students about the invention process and articles about student contest winners and their inventions. The book annotated below is an excellent resource for educators and students wanting to learn more about these contests.

Sobey, Edwin J. *How to Enter and Win an Invention Contest*. Berkeley Heights, N.J.: Enslow, 1999. 104 pp. (hc. 0-7660-1173-9) $20.95.
This book provides resources for students wanting to enter invention contests, including information on how to enter a contest, how to invent, planning the project, and a list of contests organized by state. Resources for learning more and an index conclude the book.

Camp Invention
Inventure Place
221 South Broadway Street
Akron, OH 44308-1505
Phone: 800-968-4332
FAX: 330-762-6313
Email: campinvention@invent.org
http://www.invent.org/camp_invention/
Elementary school children can participate in one-week camps held in many states. The website contains information on the dates, locations, activities, registration, and costs.

Christopher Columbus Award
(Formerly the Bayer/NSF Award for Community Innovation)
105 Terry Drive, Suite 120
Newton, PA 18940-3425
Phone: 800-291-6020
FAX: 215-579-8589
http://www.nsf.gov/od/lpa/events/bayernsf/start.htm
This competition is a collaboration between the Christopher Columbus Fellowship Foundation and the National Science Foundation (NSF). The focus is on finding and solving community problems using the scientific process. Students in grades six through eight work in teams of three or four to solve a community problem using science and technology.

Craftsman/NSTA Young Inventors Awards Program
National Science Teachers Association
1840 Wilson Boulevard
Arlington, VA 22201-3000
Phone: 888-494-4994
http://www.nsta.org/programs.craftsman.htm
This contest is for students in grades two through eight living in the United States or a United States territory. Combining science, technology, and their creativity, students invent or modify a tool.

Duracell/NSTA Scholarship Competition
National Science Teachers Association
1840 Wilson Boulevard
Arlington, VA 22201-3000
Phone: 888-255-4242

FAX: 703-243-7177
http://www.nsta.org/programs/Duracell.htm
Sponsored by Duracell through the National Science Teachers Association (NSTA), this contest is open to students in grades six through twelve. Working individually or with a partner, students design and build a machine that uses Duracell batteries.

Intel Science Talent Search (STS)
1719 N Street, NW
Washington, DC 20036
Phone: 202-785-2255
http://www.intel.com/education/sts/
The Intel Science Talent Search is a science competition that provides high school seniors an opportunity to complete a research project and have it reviewed by a national jury of professional scientists.

Invent America!
PO Box 26065
Alexandria, VA 22313
Phone: 703-942-7121
FAX: 703-461-0068
Email: inventamerica@aol.com
http://www.inventamerica.com/
This is a competition for students in kindergarten through eighth grade and is sponsored by the U.S. Patent Model Foundation. Materials for educators and students can be ordered from the website.

The National Engineering Design Challenge (NEDC)
Junior Engineering Technical Society (JETS)
1420 King Street, Suite 405
Alexandria, VA 22314-2794
Phone: 703-548-5387
FAX: 703-548-0769
http://www.jets.org/programs/nedc.cfm
The NEDC is open to students in grades nine through twelve. This competition involves teams of four students who solve an engineering-based problem by designing, building, and demonstrating a working model of an innovative product.

Kids Invent Toys
1662 East Fox Glen Avenue
Fresno, CA 93720
Phone: 866-548-5437/559-434-3046
FAX: 559-278-5914
Email: info@kidsinvent.org
http://www.kidsinvent.org/
Kids Invent! offers curriculum for one-week summer camps, after school programs, and classrooms that involves students in creative thinking, inventing, and marketing their inventions. These materials are appropriate for elementary and middle school students.

Odyssey of the Mind Program
c/o Creative Competitions, Inc.
1325 Rt. 130 South, Suite F
Gloucester City, NJ 08030
Phone: 856-456-7776
FAX: 856-456-7008
http://www.odysseyofthemind.com/
This worldwide program encourages students to think creatively and divergently as they solve open-ended problems. Students from kindergarten to college are eligible to participate in creative-thinking contests sponsored by Odyssey of the Mind.

Science Olympiad
5955 Little Pine Lane
Rochester, MI 48306
Phone: 248-651-4013
FAX: 248-651-7835
http://www.soinc.org/
Elementary, middle, and high school students work in teams to solve specific engineering and science problems. These problems may require students to invent something in order to solve the problem.

Toshiba/NSTA ExploraVision Awards
National Science Teachers Association
1840 Wilson Boulevard
Arlington, VA 22201-3000
Phone: 800-397-5679
FAX: 703-243-7177
http://www.toshiba.com/tai/exploravision/index3.html

Toshiba Corporation sponsors this contest for United States and Canadian students in kindergarten through twelfth grade. Working in research-and-development teams with the aid of an adult mentor, the children combine imagination and science to predict the future development of a certain technology.

References

Kivenson, Gilbert. *The Art and Science of Invention,* 2nd ed. Melbourne, Victoria: Van Nostrand Reinhold, 1982.

Microsoft PowerPoint. Redmond, Wash.: Microsoft Corporation.

Timeliner 5.0. Watertown, Mass.: Tom Snyder Productions.

Appendix E

Building a Core Library Collection

The following list of books is a suggestion for a core collection on Discoveries and Inventions for a school library. It is arranged in chapter order with grade levels following each title. Each title is followed by at least one source of a review by an acceptable journal.

Key to Journal Abbreviations

APPR = Appraisal
AS = American Scientist
BCCB = Bulletin of the Center for Children's Books
BL = Booklist
BR = Book Report
HB = The Horn Book Magazine
LJ = Library Journal
PW = Publisher's Weekly
RT = Reading Teacher
SA = School Arts
SBF = Science Books and Films
SLJ = School Library Journal
SM = Smithsonian

Elementary School (P–2)

Hoban, Tana. *Construction Zone*. Gr. P–1.
   BL 93 (Apr. 1, 1997): 1335.
Fowler, Allan. *It Could Still be a Robot*. Gr. P–2.
   BL 94 (Dec. 1, 1997): 633.
——. *Simple Machines*. Gr. P–2.
   BL 97 (July 2001): 2023.
Hunter, Ryan Ann. *Dig a Tunnel*. Gr. P–2.
   BL 95 (Mar. 15, 1999): 1331; SLJ 45 (Apr. 1999): 114.
Konigsburg, E. L. *Samuel Todd's Book of Great Inventions*. Gr. P–2.
   HB 68 (Jan. 1992): 59; SLJ 37 (Oct. 1991): 98.
Eck, Michael. *The Internet: Inside and Out*. Gr. P–3.
   BR 21 (Nov./Dec. 2002): 54.
Hopkinson, Deborah. *Maria's Comet*. Gr. P–3.
   BL 96 (Sept. 15, 1999): 268.
Hill, Lee Sullivan. *Dams Give Us Power*. Gr. K–2.
   BL 94 (Dec. 15, 1997): 237.
Graham, Ian. *The Best Book of Spaceships*. Gr. K–3.
   SLJ 45 (Mar. 1999): 194.
Howard, Ginger. *William's House*. Gr. K–3.
   BL 97 (Mar. 15, 2001): 1400; SLJ 47 (Mar. 2001): 212.
Carlson, Laurie. *Boss of the Plains: The Hat That Won the West*. Gr. K–4.
   BL 94 (Mar. 1, 1998): 1138; HB 74 (May/June 1998): 356.
Wells, Rosemary, and Tom Wells. *The House in the Mail*. Gr. K–4.
   BL 98 (Mar. 1, 2002): 1137; SLJ 48 (Mar. 2002): 205.
Towle, Wendy. *The Real McCoy: The Life of an African American Inventor*. Gr. K–4.
   SLJ 39 (Apr. 1993): 115.
Goldsmith, Howard. *Thomas Edison to the Rescue*. Gr. 1–2.
   SLJ 49 (Mar. 2003): 217.
Ogren, Cathy Stefanec. *The Adventures of Archie Featherspoon*. Gr. 1–3.
   SLJ 48 (Aug. 2002): 164.
Schulz, Walter A. *Will and Orv*. Gr. 1–3.
   SLJ 38 (Feb. 1992): 84.
Adler, David A. *A Picture Book of George Washington Carver*. Gr. 1–4.
   BL 95 (Apr. 15, 1999): 1532.
Brill, Marlene Targ. *Margaret Knight: Girl Inventor*. Gr. 1–4.
   SLJ 47 (Dec. 2001): 118.

Harper, Charise Mericle. *Imaginative Inventions: The Who, What, Where, When and Why of Roller Skates, Potato Chips, Marbles, and Pie.* Gr. 1–4.
SLJ 47 (Oct. 2001): 140.

Schanzer, Rosalyn. *How Ben Franklin Stole the Lightning.* Gr. 1–4.
SLJ 49 (Jan. 2003): 129.

Yolen, Jane. *My Brothers' Flying Machine.* Gr. 1–4.
BL 99 (Mar. 1, 2003): 11208.

Thomas, Mark. *The Akashi Kaikyo Bridge: World's Longest Bridge.* Gr. 2–3.
SLJ 48 (Aug. 2002): 181.

Krensky, Stephen. *Taking Flight: The Story of the Wright Brothers.* Gr. 2–4.
BL 96 (May 15, 2000): 1740.

Halperin, Wendy Anderson. *Once Upon a Company.* Gr. 2–5.
SLJ 44 (Sept. 1998): 194.

Maestro, Betsy. *The Story of Clocks and Calendars: Marking a Millennium.* Gr. 2–5.
BL 95 (June 1999): 1822.

O'Brien, Patrick. *The Hindenburg.* Gr. 2–5.
HB 76 (Sept./Oct. 2000): 598.

Old, Wendie. *To Fly: The Story of the Wright Brothers.* Gr. 2–5.
BCCB 56 (Nov. 2002): 121; HB 78 (Nov./Dec. 2002): 779–80.

Elementary School (3–5)

Wishinsky, Frieda. *What's the Matter with Albert?* Gr. 3–4.
SLJ 48 (Nov. 2002): 151.

Busby, Peter. *First to Fly: How Wilbur and Orville Wright Invented the Airplane.* Gr. 3–5.
LJ 49 (June 2003): 83.

Davis, Lucile. *Medicine in the American West.* Gr. 3–5.
SLJ 47 (Nov. 2001): 172.

Merbreier, W. Carter with Linda Capus Riley. *Television: What's Behind What You See.* Gr. 3–5.
SLJ 42 (Mar. 1996): 212.

Weitzman, David. *Jenny: The Airplane that Taught America to Fly.* Gr. 3–5.
SLJ 49 (Feb. 2003): 139.

Curlee, Lynn. *Brooklyn Bridge*. Gr. 3–6.
  SLJ 48 (Mar. 2002): 49.
Haseley, Dennis. *The Amazing Thinking Machine*. Gr. 3–6.
  HB 78 (July/Aug. 2002): 461.
Matthews, Tom L. *Always Inventing: A Photobiography of Alexander Graham Bell*. Gr. 3–6.
  BL 95 (Oct. 15, 1998): 418; SLJ 44 (Dec. 1998): 110.
Romanek, Trudee. *The Technology Book for Girls and Other Advanced Beings*. Gr. 3–6.
  BR 20 (Nov./Dec. 2001): 74.
Jones, Charlotte Foltz. *Accidents May Happen: Fifty Inventions Discovered by Mistake*. Gr. 3–8.
  BL 92 (June 1996): 1710.
Angliss, Sarah, and Colin Uttley. *Cities in the Sky: A Beginner's Guide to Living in Space*. Gr. 4–6.
  SLJ 45 (Apr. 1999): 144.
Couper, Heather, and Nigel Henbest. *Black Holes*. Gr. 4–6.
  BL 92 (Aug. 1996): 1892.
Giblin, James Cross. *The Mystery of the Mammoth Bones and How It Was Solved*. Gr. 4–6.
  HB 75 (Sept./Oct. 1999): 624.
Hargrove, Jim. *Dr. An Wang: Computer Pioneer*. Gr. 4–6.
  APPR 26 (Fall 1993): 27; SBF 30 (Mar. 1994): 53.
Kent, Peter. *Great Buildings: Stories of the Past*. Gr. 4–6.
  SA 101 (Apr. 2002): 58.
Ross, Michael Elsohn. *Toy Lab*. Gr. 4–6.
  SLJ 49 (Feb. 2003): 168.
Weidt, Maryann N. *Mr. Blue Jeans: A Story about Levi Strauss*. Gr. 4–6.
  BL 87 (Dec.1, 1990): 738; SLJ 37 (Jan. 1991): 109.
Weitzman, David. *Old Ironsides*. Gr. 4–6.
  BL 93 (May 1, 1997): 1498.
Adkins, Jan. *Bridges: From My Side to Yours*. Gr. 4–8.
  BR 21 (Nov./Dec. 2002): 68.
Bial, Raymond. *The Houses*. Gr. 4–8.
  RT 47 (Mar. 1994): 490.
Biel, Jackie. *Video*. Gr. 4–8.
  BL 92 (July 1996): 1832.
Carlson, Laurie. *Queen of Inventions: How the Sewing Machine Changed the World*. Gr. 4–8.
  BL 99 (Feb. 15, 2003): SLJ 49 (Apr. 2003): 148.

Casey, Susan. *Women Invent!: Two Centuries of Discoveries That Have Shaped Our World*. Gr. 4–8.
    PW 244 (Nov. 17, 1997): 63.
Duffy, Trent. *The Clock*. Gr. 4–8.
    BR 19 (Nov./Dec. 2000): 69; SLJ 46 (May 2000): 180.
Fox, Mary Virginia. *Edwin Hubble: American Astronomer*. Gr. 4–8.
    BL 94 (Dec. 1, 1997): 621; SLJ 43 (Nov. 1997): 127.
Gifford, Clive. *How to Build a Robot*. Gr. 4–8.
    SLJ 48 (June 2002): 158.
Gourley, Catherine. *Good Girl Work: Factories, Sweatshops, and How Women Changed Their Role in the American Workforce*. Gr. 4–8.
    BL 95 (May 1, 1999): 1584.
Ichord, Loretta Frances. *Toothworms and Spider Juice: An Illustrated History of Dentistry*. Gr. 4–8.
    BL 96 (Feb. 15, 2000): 1108.
Martin, Jacqueline Briggs. *Snowflake Bentley*. Gr. 4–8.
    HB 74 (Sept./Oct. 1998): 622.
Matthews, Tom L. *Light Shining through the Mist: A Photobiography of Dian Fossey*. Gr. 4–8.
    BL 95 (Sept. 15, 1998): 223; HB 74 (Sept./Oct. 1998): 622; SLJ 44 (Sept. 1998): 222.
Morgan, Nina. *Lasers*. Gr. 4–8.
    BL 93 (June 1997): 1696.
Murdico, Suzanne J. *Concorde*. Gr. 4–8.
    BR 20 (Nov./Dec. 2001): 84.
O'Brien, Patrick. *Duel of the Ironclads: The Monitor vs. the Virginia*. Gr. 4–8.
    BL 99 (Mar. 15, 2003): 1323.
Richardson, Hazel. *How to Clone a Sheep*. Gr. 4–8.
    SLJ 48 (June 2002): 158.
Riehecky, Janet. *Television*. Gr. 4–8.
    SLJ 42 Aug. 1996): 148.
Rinard, Judith E. *The Story of Flight: The Smithsonian National Air and Space Museum*. Gr. 4–8.
    SLJ 49 (June 2003): 83.
Rubin, Susan Goldman. *There Goes the Neighborhood:Ten Buildings People Loved to Hate*. Gr. 4–8.
    SLJ 48 (Mar. 2002): 49.
Wilsdon, Christina. *Everyday Things: An A–Z Guide*. Gr. 4–8.
    BR 20 (Mar./Apr. 2002): 73.

Tchudi, Stephen. *Lock and Key: The Secrets of Locking Things Up, In and Out.* Gr. 4–12.
    BL 90 (Dec. 1, 1993): 689; BR 12 (Mar./Apr. 1994): 49.
Wallace, Joseph. *The Camera.* Gr. 4–12.
    SLJ 47 (Jan. 2001): 148.
_____. *The Lightbulb.* Gr. 4–12.
    SLJ 45 (Oct. 1999): 168.
Aaseng, Nathan. *Construction: Building the Impossible.* Gr. 5–8.
    BR 19 (May/June 2000): 68.
Anderson, Margaret J. *Charles Darwin: Naturalist.* Gr. 5–8.
    APPR 27 (Fall 1994): 67; BL 91 (Jan. 1, 1995): 816;
    SBF 31 (Jan./Feb. 1995): 24; SLJ 40 (Oct. 1994): 129.
Baker, Christopher W. *Robots among Us: The Challenges and Promises of Robotics.* Gr. 5–8.
    BL 98 (June 1, 2002): 1710.
Barron, Rachel Stiffler. *Lise Meitner: Discoverer of Nuclear Fission.* Gr. 5–8.
    BL 96 (Mar. 15, 2000): 1368; SLJ 45 (June 2000): 158.
Byman, Jeremy. *Carl Sagan: In Contact with the Cosmos.* Gr. 5–8.
    BL 97 (Nov. 1, 2000): 528; SLJ 46 (Aug. 2000): 196.
Datnow, Claire. *Edwin Hubble: Discoverer of Galaxies.* Gr. 5–8.
    BL 94 (Dec. 1, 1997): 630; SLJ 44 (Mar. 1998): 230.
Donnellan, William L. *The Miracle of Immunity.* Gr. 5–8.
    SLJ 49 (May 2003): 166.
Erlbach, Arlene. *The Kids' Invention Book.* Gr. 5–8.
    BL 94 (Feb. 1, 1998): 912.
Klare, Roger. *Gregor Mendel: Father of Genetics.* Gr. 5–8.
    BL 94 (Dec. 1, 1997): 625; SBF 34 (Jan./Feb. 1998): 23; SLJ 43 (Dec. 1997): 139.
Mulcahy, Robert. *Medical Technology: Inventing the Instruments.* Gr. 5–8.
    SLJ 43 (July 1997): 109.
Pflaum, Rosalynd. *Marie Curie and Her Daughter Irene.* Gr. 5–8.
    BL 89 (Aug. 1993): 2048; SBF 29 (June/July 1993): 145; SLJ 39 (June 1993): 122.
Rubin, Susan Goldman. *Toilets, Toasters, and Telephones.* Gr. 5–8.
    BL 95 (Nov.1, 1998): 488; HB 75 (Mar./Apr. 1999): 29.
Streissguth, Thomas. *Communications: Sending the Message.* Gr. 5–8.
    SLJ 44 (Feb. 1998): 127.
Vanderwarker, Peter. *The Big Dig: Reshaping an American City.* Gr. 5–8.
    HB 78 (Jan./Feb. 2002): 107.

Williams, Brian. *Computers*. Gr. 5–8.
SLJ 48 (Feb. 2002): 142.

Yannuzzi, Della A. *Madam C. J. Walker: Self-Made Businesswoman*.
Gr. 5–8.
BL 96 (Feb. 15, 2000): 1102; SLJ 46 (July 2000): 126.

Marrin, Albert. *Dr. Jenner and the Speckled Monster: The Search for the Smallpox Vaccine*. Gr. 5–9.
HB 78 (Nov./Dec. 2002): 777.

French, Laura. *Internet Pioneers: The Cyber Elite*. Gr. 5–12.
SLJ 47 (Sept. 2001): 242.

Pringle, Laurence. *The Environmental Movement*. Gr. 5–12.
SLJ 149 (Aug. 2003): 116.

Sherman, Josepha. *Jerry Yang and David Filo: Chief Yahoos of Yahoo!*
Gr. 5–12.
SLJ 47 (Dec. 2001): 153.

Middle School (6–8)

Parker, Steve. *Fuels for the Future*. Gr. 6–8.
BR 17 (Sept./Oct. 1998): 66.

Cobb, Allan B. *How Do We Know How Stars Shine*. Gr. 6–9.
BR 21 (May/June 2002): 72

DuTemple, Lesley A. *The New York Subways*. Gr. 6–9.
BL 99 (Jan. 2003): 878.

Goldstein, Natalie. *How Do We Know the Nature of the Atom*. Gr. 6–9.
SLJ 47 (Dec. 2001): 154.

Skinner, David. *The Wrecker*. Gr. 6–9.
SLJ 41 (July 1995): 96.

Yount, Lisa. *History of Medicine*. Gr. 6–9.
SLJ 48 (Jan. 2002): 142.

Conley, Kevin. *Benjamin Banneker: Scientist and Mathematician*.
Gr. 6–10.
APPR 23 (Summer 1990): 57–58; BL 86 (Jan.1, 1990): 912; SLJ
36 (May 1990): 130.

Altman, Linda Jacobs. *Women Inventors*. Gr. 6–12.
BR 16 (Sept./Oct. 1997): 46.

Collier, Bruce, and James MacLachlan. *Babbage and the Engines of Perfection*. Gr. 6–12.
BL 95 (Dec. I, 1998): 674; SLJ 45 (Feb. 1999): 114.

Davies, Paul. *How to Build a Time Machine*. Gr. 6–12.
    BL 98 (Feb. 1, 2002): 912.
Hager, Tom. *Linus Pauling and the Chemistry of Life*. Gr. 6–12.
    BL 94 (May 15, 1998): 1614; SBF 34 (Oct. 1998): 207; SLJ 44
    (Aug. 1998): 174.
Karnes, Frances A., and Suzanne M. Bean. *Girls and Young Women
    Inventing: Twenty True Stories about Inventors Plus How You Can
    Be One Yourself*. Gr. 6–12.
    BL 92 (Feb. 1, 1996): 924; SLJ 41 (Dec. 19, 1995): 119.
Karwatka, Dennis. *Technology's Past: America's Industrial Revolution
    and the People Who Delivered the Goods*. Gr. 6–12.
    BR 15 (Nov./Dec. 1996): 55.
Macaulay, David. *Cathedral: The Story of Its Construction*. Gr. 6–12.
    BL 96 (Nov. 15, 1999): 620.
Pasachoff, Naomi. *Marie Curie and the Science of Radioactivity*. Gr. 6–12.
    BL 93 (Sept. 1, 1996): 115; SLJ 42 (Aug. 1996): 174.
Graham, Ian. *Internet Revolution*. Gr. 7–10.
    SLJ 49 (Jan. 2003): 160.
Billings, Charlene W. *Grace Hopper: Navy Admiral and Computer
    Pioneer*. Gr. 7–12.
    APPR 23 (Spring 1990): 20; BCCB 43 (Dec. 1989): 79; BL 86
    (Oct. 15, 1989): 436; SLJ 36 (Mar. 1990): 242.
Bridgman, Roger. *1000 Inventions and Discoveries*. Gr. 7–12.
    SLJ 49 (Mar. 2003): 246.
Cooney, Caroline. *Both Sides of Time*. Gr. 7–12.
    BR 14 (May/June 1995): 37.
_____. *Out of Time*. Gr. 7–12.
    SLJ 49 (May 2003): 50.
Cross, Gillian. *New World*. Gr. 7–12.
    HB 71 (July/Aug. 1995): 465.
Dunn, John M. *The Computer Revolution*. Gr. 7–12.
    SLJ 48 (Apr. 2002): 170.
St. George, Judith. *Dear Mr. Bell . . . Your Friend, Helen Keller*. Gr. 7–12.
    BCCB 46 (Feb. 1993): 192; HB 68 (Nov./Dec. 1992): 736;
    SLJ 38 (Dec. 1992): 129.
Sherrow, Victoria. *Jonas Salk*. Gr. 7–12.
    BL 90 (Sept. 15, 1993): 143; SBF 29 (Nov. 1993): 236; SLJ 39
    (Sept. 1993): 258.
Tucker, Tom. *Brainstorm!: The Stories of Twenty American Kid In-
    ventors*. Gr. 7–12.
    BL 91 (Aug. 1995): 1945; BR 14 (Jan./Feb. 1996): 60.

Fridell, Ron. *Solving Crimes: Pioneers of Forensic Science.* Gr. 8–12.
BR 19 (Jan./Feb. 2001): 80.

Levy, Joel. *Really Useful: The Origins of Everyday Things.* Gr. 8–12.
BL 99 (Jan. 2003): 824.

High School (9–12)

Bortz, Fred. *Techno-Matter: The Materials behind the Marvels.* Gr. 9–12.
SLJ 47 (Apr. 15, 2001): 179.

Brain, Marshall. *Marshall Brain's How Stuff Works.* Gr. 9–12.
SLJ 48 (Winter 2002): 48.

Brockman, John, ed. *The Greatest Inventions of the Past Two
Thousand Years.* Gr. 9–12.
BL 96 (Dec. 1, 1999): 672.

Geison, Gerald L. *The Private Science of Louis Pasteur.* Gr. 9–12.
BL 91 (Apr. 1, 1995): 1367.

Hellman, Hal. *Great Feuds in Medicine: Ten of the Liveliest Disputes
Ever.* Gr. 9–12.
AS 89 (July/Aug. 2001): 374.

Macaulay, David. *The New Way Things Work.* Gr. 9–12.
BL 95 (Dec. 1, 1998): 674.

Menzel, Peter, and Faith D'Aluisio. *Robo Sapiens: Evolution of a New
Species.* Gr. 9–12.
PW 247 (July 3, 2000): 55.

Perlin, John. *From Space to Earth: The Story of Solar Energy.* Gr. 9–12
LJ 124 (Oct. 15, 1999): 102.

Schwartz, Evan I. *The Last Lone Inventor: A Tale of Genius, Deceit,
and the Birth of Television.* Gr. 9–12.
LJ 127 (May 15, 2002): 100.

Standage, Tom. *The Victorian Internet: The Remarkable Story of the
Telegraph and the Nineteenth Century's On-line Pioneers.* Gr. 9–12.
LJ 123 (Sept. 15, 1998): 109.

Vare, Ethlie Ann, and Greg Ptacek. *Patently Female: From AZT to TV
Dinners: Stories of Women Inventors and Their Breakthrough
Ideas.* Gr. 9–12.
BL 98 (Dec. 1, 2001): 619.

Brands, H. W. *The First American: The Life and Times of Benjamin
Franklin.* Gr. 10–12.
LJ 125 (Sept. 15, 2000): 85.

Breuer, William B. *Secret Weapons of World War II*. Gr. 10–12.
    PW 247 (Sept. 18, 2000): 102.
Feldman, Burton. *The Nobel Prize: A History of Genius, Controversy,
    and Prestige*. Gr. 10–12.
    BL 97 (Dec. 1, 2000): 677.
Friedel, Robert. *Zipper: An Exploration in Novelty*. Gr. 10–12.
    LJ 119 (Mar. 15, 1994): 98.
Greenstein, George. *Portraits of Discovery: Profiles in Scientific
    Genius*. Gr. 10–12.
    SM 30 (Aug. 1999): 123.
Israel, Paul. *Edison: A Life of Invention*. Gr. 10–12.
    BL 95 (Oct. 1, 1998): 299; LJ 123 (Oct. 15, 1998): 299.
Lindsay, David. *House of Invention: The Secret Life of Everyday
    Products*. Gr. 10–12.
    LJ 125 (Mar. 15, 2000): 123.
Ne'eman, Yuval, and Yoram Kirsh. *The Particle Hunters*, 2nd ed.
    Gr. 10–12.
    AS 76 (May/June 1988): 297.
O'Connell, Robert L. *Soul of the Sword: An Illustrated History of
    Weaponry and Warfare from Prehistory to the Present*. Gr. 10–12.
    LJ 127 (Sept. 1, 2992): 98.
Slack, Charles. *Noble Obsession: Charles Goodyear, Thomas Hancock,
    and the Race to Unlock the Greatest Industrial Secret of the
    Nineteenth Century*. Gr. 10–12.
    BL 98 (July 1, 2002): 1804.
Wherry, Timothy Lee. *Patent Searching for Librarians and Inventors*.
    Gr. 10–12.
    LJ 120 (June 15, 1995): 100.
Tenner, Edward. *Why Things Bite Back: Technology and the Revenge
    of Unintended Consequences*. Gr. 11–12.
    LJ 121 (May 15, 1996): 82.

Author Index

Title Index

Subject Index

acquired immune deficiency syndrome, 14, 128–130
aerospace, 156
aetherphone, 231
African American inventors, 1, 2, 9, 37, 57, 60, 62, 63, 73
African American physicians, 14, 35, 59
African American scientists, 5, 7, 11, 13, 42, 57, 60, 62, 63, 65, 66, 68, 74, 75, 149, 150, 154, 161, 202, 218, 242, 245, 293, 295
AIDS. *See* acquired immune deficiency syndrome
airplane, 3, 7, 22, 34, 62, 70, 151, 153, 157, 177, 178, 236, 291
airship, 152, 156
Alaska Pipeline, 98
Allen, Paul, 26
Al-Khwarizmi, 61
alloy, 110, 225
alphabet, 11, 120, 122
Alvarez, Luis, 63, 66
anatomy, 130
ancient inventions, 237
Anderson, Mary, 174
anesthesia, 2, 130. *See* ether
Andrews, Roy Chapman, 24
Anning, Mary, 28, 71, 72, 74, 76
anthrax, 44, 162, 232, 280
antibiotic, 36, 41, 130
archeologist, 57, 59
Archimedes, 65, 177
architecture, 100, 203
Aristotle, 14, 65
Armstrong, Edwin Howard, 61
artificial heart, 20
artificial intelligence, 3, 53, 90, 123, 187, 189, 192, 248

Asian American scientists, 2, 64
aspirin, 172, 274
astronaut, 143, 144, 156
astronomer, 7, 16, 21, 27, 30–33, 39, 42, 53, 57, 61, 70, 142, 143, 208, 250, 254, 262, 293
astronomy, 30, 53, 59, 61, 146, 195, 196, 198–200, 208, 221, 222, 243, 251, 254, 269
AstroTurf, 104
atom, 32, 49, 133, 136, 137, 139, 173, 200, 250, 295
automation, 173
atomic, 34, 91, 92, 205, 238, 250
Atomic Age, 15, 78
atomic bomb, 46, 49, 161, 163, 164, 220, 241, 270
automobile, 3, 13, 33, 42, 68, 74, 151–153, 156, 173, 184, 195, 196, 228
Avery, Oswald, 25
aviation, 14, 17, 68, 153, 156, 157, 218, 223, 230, 235, 251, 264, 265
Aztec, 170

Babbage, Charles, 38, 44, 238, 295
Baby Ruth, 103
bacteria, 36, 37, 41, 130
Baird, John Logie, 40
Baker, S. Josephine, 18, 63, 66
Ballard, Geoffrey, 241
Band-Aid, 100
Banneker, Benjamin, 7, 13, 42, 63, 202, 295
Banting, Frederick, 19, 60, 244
bar-code, 230
Barnard, Christiaan, 19
battleship, 152, 162

About the Authors

Joy L. Lowe was born in Minden, Louisiana. She received undergraduate and graduate degrees from Centenary College of Louisiana, Louisiana Tech University, and Louisiana State University. She received a Ph.D. in library and information science from the University of North Texas. A former school and public librarian, she taught library science at Louisiana Tech University for twenty-six years. Joy coauthored the *Colonial America in Literature for Youth: A Resource Guide* and *Neal-Schuman Guide to Recommended Children's Books and Media for Use with Every Elementary Subject*. She lives in Ruston, Louisiana, with her husband, Perry.

Kathryn I. Matthew was born in Oakland, California. She received an Ed.D. in curriculum and instruction with an emphasis on technology and reading from the University of Houston. She has worked in elementary schools in Texas and Louisiana as a classroom teacher, English as a second language specialist, and technology specialist. At the university level she has taught children's literature, reading, language arts, technology, and research classes. Kathryn co-authored *Colonial America in Literature for Youth: A Resource Guide*; *Reading Comprehension: Books and Strategies for the Elementary Curriculum*; *Technology, Reading and Language Arts*; and the *Neal-Schuman Guide to Recommended Children's Books and Media for Use with Every Elementary Subject*. She and her husband, Chip, live in Sugar Land, Texas.